A HISTORY OF THE KENT ISLAND volunteer FIRE DEPARTMENT

A HISTORY OF THE KENT ISLAND VOLUNTEER FIRE DEPARTMENT

BRENT LEWIS

THE History PRESS

Published by The History Press
Charleston, SC 29403
www.historypress.net

Front cover, from top left, clockwise: *Courtesy Joyce Moore, AsiPhotos.com/Stan Ruddie, Tommy Norman and Will Clark.*

Back cover: *Courtesy AsiPhotos.com/Stan Ruddie.*

First published 2010

Manufactured in the United States

ISBN 978.1.59629.981.8

Library of Congress Cataloging-in-Publication Data

Lewis, Brent.
A history of the Kent Island Volunteer Fire Department / Brent Lewis.
p. cm.
Includes bibliographical references.
ISBN 978-1-59629-981-8
1. Kent Island Volunteer Fire Dept. (Kent Island, Md.)--History. 2. Kent Island Volunteer Fire Dept. (Kent Island, Md.)--History--Pictorial works. 3. Fire fighters--Maryland--Kent Island--Biography. 4. Kent Island Volunteer Fire Dept. (Kent Island, Md.)--Biography. 5. Fire extinction--Maryland--Kent Island--History. 6. Kent Island (Md.)--History. 7. Kent Island (Md.)--Biography. 8. Kent Island (Md.)--Social life and customs. I. Title.
TH9504.M374L49 2010
363.3709752'34--dc22
2010029501

CONTENTS

Contents

PREFACE

When my mom was a little girl, her parents operated a combination gas station, store and sandwich shop on Kent Island's Romancoke Road, across from Camp Wright at the tip of one of Warehouse Creek's marshy fingers. They catered to motorists using the Matapeake Ferry and lived in an apartment upstairs. One afternoon, probably in 1940, the family of three—my aunt Carolyn wasn't born yet—went to visit family off-island. While they were gone, everything they owned went up in flames.

Even if you've never personally experienced the tragedy of losing all your possessions—or even worse, loved ones—to fire or accident, there is an awareness of that potential horror ingrained in all of us. Everybody's life, either intimately or through association, can be touched by misfortune and catastrophe.

Sometimes life is harsh.

Yet there are those among us who volunteer their time, their efforts and even their lives to making the rest of us safer. They provide us with the chance, improve the odds, that maybe the worst moments of our existence might turn out okay after all. They make life a little less harsh.

This book is meant to provide a short narrative history of Kent Island's last six decades through the eyes of some of the people in our community who do just that. And although this is the story of one specific volunteer fire department, it's also meant to reflect the compassion, bravery, dedication and kinship universal in the mettle of those who serve.

Much of this history of the Kent Island Volunteer Fire Department is based on oral history interviews conducted from 2008 to 2010. Where

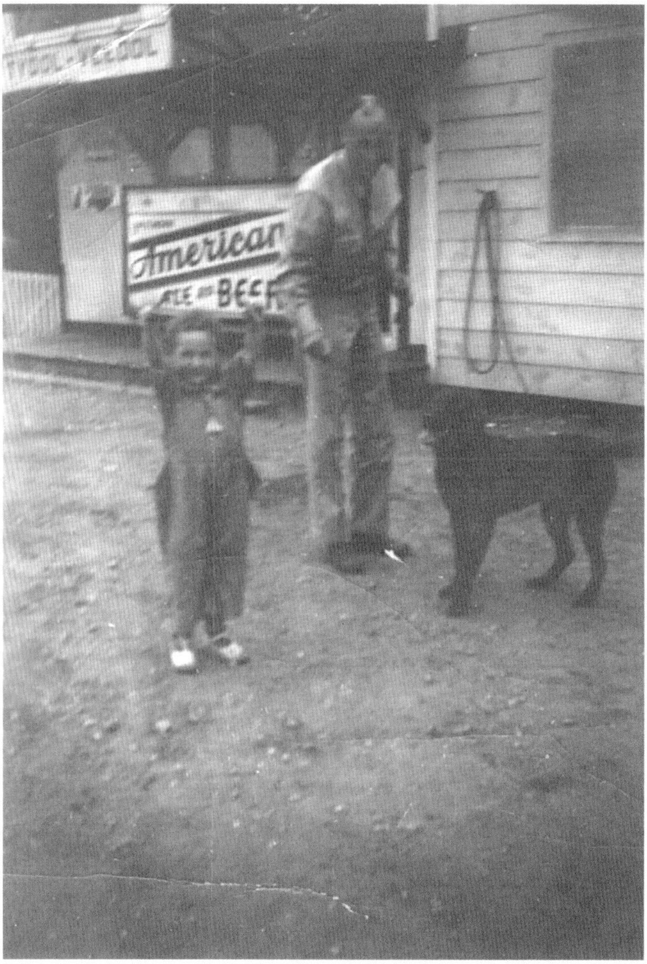

My mother, Jackie, with her dad, Grover Clough, and beloved Chesapeake Bay retriever, Brownie, at their Romancoke Road property, later destroyed by fire. *Author's collection*

possible, details have been corroborated, but it is acknowledged that oral histories are anecdotal and leave room for disagreement in points of view.

My participation in this project is dedicated to first responders everywhere and to previous and subsequent generations of my family and yours.

ACKNOWLEDGEMENTS

Sincere thanks to the hundreds of unsung Kent Island Volunteer Fire Department heroes whose contributions and stories are not included in this book; The Kent Island Heritage Society membership for their support and continued dedication "To Discover, Identify, Restore and Preserve the Heritage of Kent Island"; everyone who provided photos and recipes; the librarians of Queen Anne's County (they're the best and always have been); local newspapers, past and present, especially the *Bay Times* and the *Queen Anne's Record-Observer* (this project would have been impossible without access to their archives); and all the journalists who covered the Kent Island Volunteer Fire Department over the decades, including Chris Rosendale, Dan Tabler, Bettye Speed, Eve Horney, Angela Price, Rose Spik, Leslie Crook, Camille O'Donnell, Bill Evans, John T. Carpozzi, David Rambo, Bill Thompson, Bill Carroll, Doug Bishop, Jack Shaum and those omitted due to oversight or lack of bylines.

I personally thank Peggy for everything; Nicole for being the apple of my eye and the chip off my block; Norma for unwavering support and encouragement; Mom for undying inspiration; Dad for occasionally not letting facts get in the way of a good story; Jody for seeing the need for this project; The History Press; John Bonner for being a top-notch first reader; Bruno and Shy for being real good dogs; all my supporters; my boys; and Ralph "I'm a Brick" Wiggum.

Visit the Kent Island Volunteer Fire Department website at www.kivfd.org.

Part I
WHERE THERE'S SMOKE

Where there's smoke, there's fire.
—common adage

Maryland's Eastern Shore, Queen Anne's County and Kent Island

Billy Lane is the sole surviving charter member of the Kent Island Volunteer Fire Department (KIVFD).

Born on September 1, 1921, on a Batts Neck Farm, William Emory Lane was the only child of Steadman and Elizabeth (Palmer) Lane. The Lanes were a typical Kent Island farm family of the early twentieth century—hardworking and of uncomplicated pleasures.

Kent Island was still a place of dirt or oyster shell roads leading to modest centers of commerce and activity, enclaves of people who worked on the water and a few scattered manors and farmhouses. Billy attended a down-island one-room schoolhouse and graduated from Stevensville High School in 1939. He married Grasonville's Muriel O'Donnell four years later. The couple raised three sons.

Billy Lane remembers Kent Island before the KIVFD was founded in 1947. When asked what happened when a Kent Islander's house caught fire in those days, Billy's answer is blunt, "It burned down."

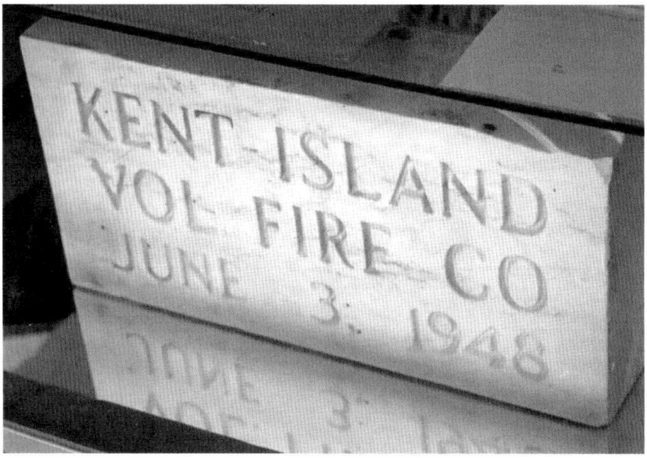

Cornerstone from Kent Island Volunteer Fire Department's original firehouse, now honorably displayed in the new Chester station's memorabilia showcase. *Courtesy AsiPhotos.com/ Stan Ruddie.*

The Chesapeake Bay just about divides Maryland in half. Both halves were colonized early in our country's history, and despite a particularly shaky political start and the occasional flare-up of simmering antagonisms, the Eastern and Western Shores have always harbored shared interests. For many decades, however, the two sides were physically separate and worlds apart. The Western Shore megalopolis of Washington, D.C., Baltimore and Annapolis was the seat of every kind of power—a growing sprawl. The Eastern Shore, made up of small towns and slightly bigger small towns, farmland and hardscrabble waterman communities, was not only the state's agricultural breadbasket but its seafood basket as well. Eastern Shorefolk proudly called their home "the Sho'" and meant it when they said they inhabited the "Land of Pleasant Living." Kent Island is part of that tradition.

Founded by the ambitious surveyor and fur trader William Claiborne in 1631, Kent Island was among America's first permanent English settlements. A bitter and long-running feud between the Virginian Claiborne and the Calverts of St. Mary's for control of the island ended in the Marylanders' favor. When Queen Anne's County was established in 1706, Kent Island was included in the new jurisdiction.

Kent Island's character and culture, like much of the Eastern Shore, still leans toward the South. Like the state, Kent Island is divided into approximate halves—from the bay to Cox Creek is Stevensville; from Cox Creek to Kent Narrows is Chester.

Historically, the Chesapeake Bay's largest island was connected to "the Other Side" by one overworked ferry or steamship system or another and to the Eastern Shore mainland via railroad. With the expansion of car ownership

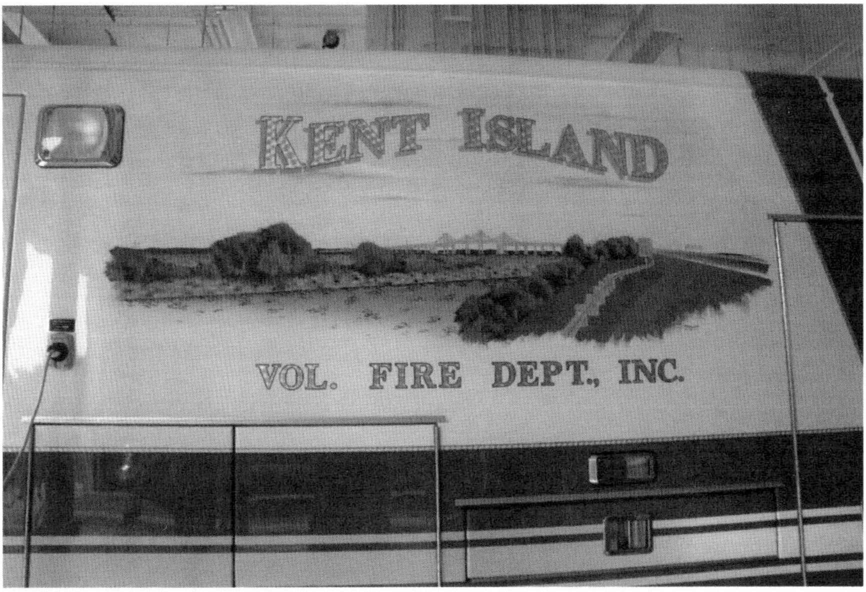

Kent Island: Gateway to the Eastern Shore. All KIVFD apparatus are proudly "Paid for by the Community We Serve." *Courtesy AsiPhotos.com/Stan Ruddie.*

and the creation of the National Highway System in the mid-twentieth century, cultural change in the region picked up pace. The Chesapeake Bay Bridge, later renamed the William Preston Lane Jr. Memorial Bay Bridge to honor the Maryland governor who championed its construction, was completed in 1952. On July 30 of that year, former Governor Lane and his successor, Theodore Mckeldin, led an inaugural procession from the western terminus at Anne Arundel County's Sandy Point to Kent Island. After that short four-mile ride, life on "the Sho'" would never be the same.

Some of our nation's story, as well as Kent Island's, can be told through flame. A 1608 Jamestown fire ignited in the middle of a grim winter and destroyed most of that already desperate colony. Kent Island's first recorded fire occurred twenty-three years later when one of Claiborne's storehouses burned down. Massive and devastating repetitive fires in Boston, New York, Charleston and New Orleans taught hard-earned lessons concerning fire prevention and extinguishment. The Great Baltimore Conflagration of 1904 crippled the Chesapeake Bay's regional economy for years.

Initially, firefighting organizations were for-profit protectors of insured property or volunteers, such as Benjamin Franklin's Union Fire Company in Philadelphia. George Washington and Thomas Jefferson, among other founding fathers, served as volunteer firefighters. Equipment employed by early first responders—leather buckets, simple ladders and hand pumps—was rudimentary at best. Rival private companies existed in many cities, and insurers stoked competition with the practice of paying whichever group put out a fire on one of their underwritten properties. On occasion, opposing firemen brawled over the right to save a blazing structure, sometimes while that structure burned to the ground.

Country folk were generally left to fight their own fires.

Over time, advances in professionalism helped the fireman reach a valued status in society. Strides made in the development of technique, as well as apparatus, gear and alarm systems, gave rise to a highly trained and better-equipped firefighter.

Centreville is the seat of Queen Anne's County. Incorporated in 1794, today the town is home to beautiful Victorian-era homes, late nineteenth-century commercial districts and the oldest continuously used courthouse in Maryland. Centreville's Goodwill Fire Company was the county's first permanent firefighting organization. Established in 1889, Goodwill is at least fifteen miles from the nearest spot on Kent Island. The county seat's volunteers were of little, if any, assistance to the people of the island. After a couple early false starts, the Queenstown Volunteer Fire Department formed in 1935. This brought help closer by about half the distance, but Kent Islanders felt change coming and realized that half the distance was still too far.

From a Little Spark

In February 2008, as the members of the Kent Island Volunteer Fire Department vacated their longtime home at Stevensville's 605 Main Street for the new state-of-the-art Chester station, some of the organization's leadership discovered a trove of historical treasure. Inside the cornerstone of their original firehouse was a Mason jar containing a Bible, five pennies, a Lions Club pin, a photograph, a history of the department's first year and a yellowed program from the building's June 3, 1948 dedication. According to that program, the ceremony's guest speaker was Baltimore's new mayor, Thomas D'Alesandro Jr., whose city had recently donated a reconditioned 1941 ambulance to the new fire company. Lions Club president T. Walter Denny and building committee mason Orem Thompson laid the cornerstone.

It's getting harder to find people who remember what Kent Island was like before that long-ago summer day.

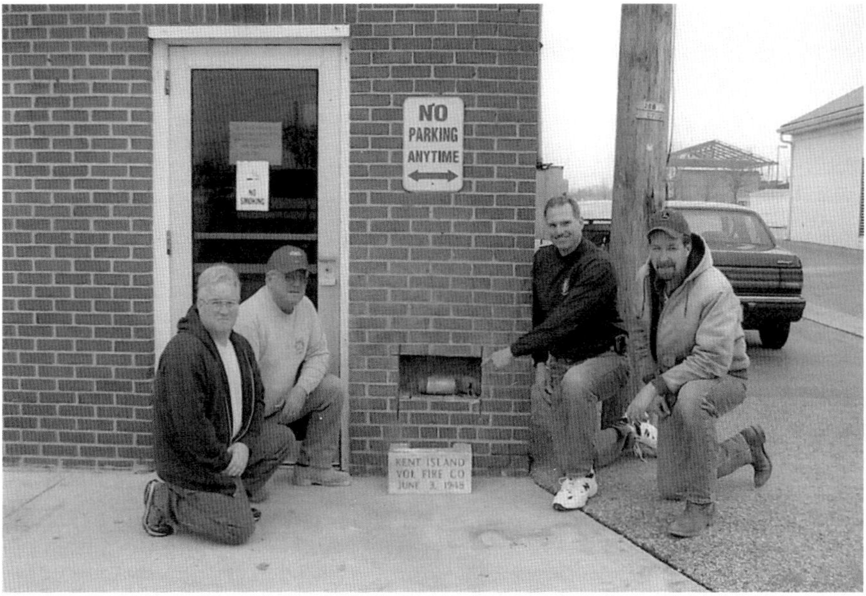

February 2008. *From left*: Paul Schlotterbeck, Lou Johnson, Jody Schulz and Buddy Thomas remove the cornerstone of the original firehouse. *Courtesy Sandra Early.*

Crafford (or Crayford) was the seventeenth-century personal estate of Kent Island's founder William Claiborne. When Billy Lane was sixteen years old, more than a decade before the establishment of the KIVFD and when from "Matapeake down was a dirt road," the chimney on the Crafford farmhouse, built in the 1880s, caught fire. No one nearby owned a phone, so Billy drove to the Matapeake ferry terminal to call Queenstown's fire department. Meanwhile,

> *Mr. Luther Baxter and his wife kept that house from burning down. They had two ladders, one up to the porch and one on the porch roof. She'd get bucketfuls of well water by hand and take them up to him on the porch roof. He'd wet down the roof. They did that until the fire department finally got there.*

Billy laughs. "That old fire truck didn't even have a cab on it."

Billy Lane was a farm boy. W. Raymond "Ray" Baker was raised in town and a second-generation Kent Island volunteer. Ray joined the KIVFD in the early 1950s. Like Billy Lane, Ray Baker recalls what life was like before the islanders formed their own emergency response services:

We lived in what's called the Legg House in Stevensville. I'd attended the senior play at the old high school and was walking home with friends. We looked down the street where I lived and saw this bright glow. We realized it was a house on fire. I thought it was mine. We got closer and could see it was the house next door. Leonard Harper, the blacksmith, lived there. Flames were coming out, rolling up maybe five feet. Our roof was wood shingle and it was steaming. Men were up on the peak of that house with buckets fighting back the fire. Everybody was running around getting their kids and pets inside. Queenstown was the closest fire company. They'd been called but it took a long while for them to get there. My mother and father were at the Queenstown movies, a rare occasion. Someone announced the fire and my parents jumped in the car. My mother drove because my father wouldn't drive fast enough. She passed the fire trucks. Come to find out one of the fire trucks ran out of gas. They had to stop at a garage and get fuel.

Retired businessman and son of a KIVFD charter member William E. "Billy" Denny III also remembers this incident:

All we had were wells and buckets to dip the water out and run them to the fire. Some of the guys ran to the pool hall to get help. I got buckets from the garage, about thirty of them. That old wooden house, everything's dry—it went up like tinder. Once it got going we had no chance of stopping it. There were a lot of barn fires, too. Green hay seems to draw lightening, and if you've ever been in a loft filled with hay it gets so hot up there it has a combustible energy.

House and barn fires weren't the only emergencies with an outcome probably affected by the island's lack of services. Alvin "Babe" Grollman is a well-known local farmer in his mid-eighties, as well as a son and brother of founding members. Babe recalls December 1936, when six-year-old first grader Karl Kolata drowned while walking home from school with a friend. Not far from the 1909 Stevensville schoolhouse was a cistern, a deep reservoir of water, to be used in case of a fire. Babe says that young Karl "went to slide on the ice. He went right through and right straight down. Clayland Clark had to get him out with a pair of oyster tongs."

A news account from the day described how "his little companion kept on uptown, where he told several people Karl had fallen in a hole." It concluded:

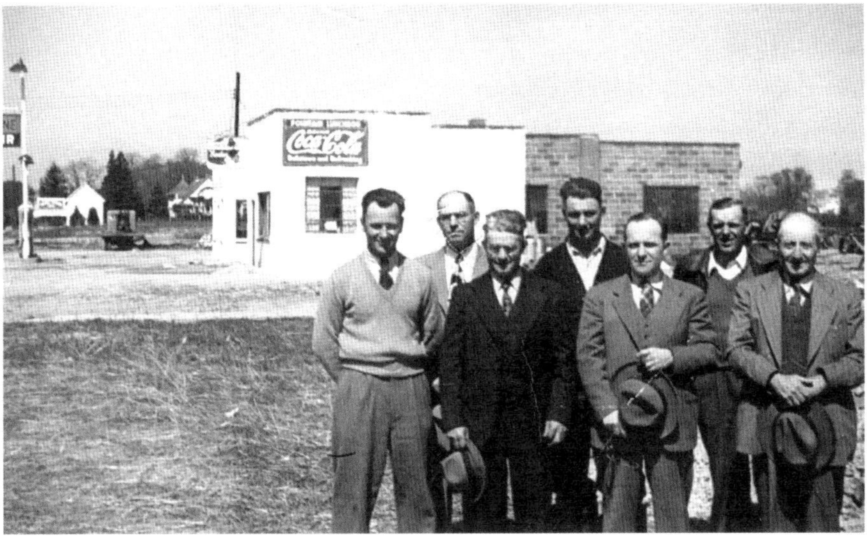

Building committee. *From left*: Orem Thompson, Dudley Sparks, Hill Hoxter, Benton Gardner, William E. Denny, Leonard Risley and Henry Grollman. Claude Lowery is not pictured. *Courtesy KIVFD.*

> *The father, Charles Kolata, who is employed by the Baltimore and Eastern Railroad, was notified, and upon going to the scene, found several residents grappling for the body. He* [Karl] *was recovered about 3:45. The little lunchbox that the child had taken to school was found nearby.*

A string of worrisome fires in the 1930s and '40s culminated in the loss of an old Stevensville structure known, long after Dr. Snyder was gone, as Dr. Snyder's place. This February 1947 incident was the catalyst for action. Though there are alternate versions of the department's origin, the official account begins in Chester when four islanders met at Myrtle Seymour's store and discussed organizing a fire department. The *Queenstown News*, one month later, acknowledged George Taylor, Leonard Risley, Joseph (Joe Buck) Baxter and Benton Gardner for their efforts to launch the fire department project.

Dolores Anthony married Kent Islander Robert "Bobby" Timms and moved to the island in 1955. Dolores has officially served in the Kent Island Volunteer Fire Department Ladies Auxiliary since 1968, acted as president for more than half a dozen years and is the department's historian. She comes from a long line of emergency volunteers. The Anthony family is known across the Eastern Shore for their commitment to various area fire companies. When Dolores was born, her father, Charles Anthony, was chief of the Queenstown Volunteer Fire Department. Early in 1947, he and three other men from Queenstown—Charles Bishop, Benny Lane and Henry Melvin—came to Kent Island to help get the new company started. Organizational meetings were held at Stevensville High School, an island landmark from 1930 to 1966.

Retired businessman Billy Denny speaks of the company's organizers: "My dad, Mr. Whitefield Coleman, Claude Lowery, George Taylor, Leonard Risley and Henry Grollman were all good friends."

Billy Lane says, "We named people to take various offices until a vote was held. Lee Bell and Henry Grollman were businessmen. Benton Gardner was a real worker. They seemed like the nucleus."

Born on March 10, 1911, Lee Bell was a graduate of Washington College in Chestertown, Maryland, and a successful Kent Island entrepreneur. His father was the Chester Station railroad agent for thirty years.

Henry Grollman was an emigrant from Lithuania. He and his wife, Lena, owned a Stevensville hardware store, where they raised their family. His son, Julius, eventually owned and expanded the store that was a Kent Island landmark for over seventy years.

Benton Gardner was a native Kent Islander with deep Eastern Shore roots. In a land where nicknames such as "Seaweed," "Dinkeye" and "Fatback" abound, Benton Gardner was called "Crust." "I guess he could be grouchy and always had a snappy comeback," speculates Billy Denny. Benton "Crust" Gardner was a pioneer guiding force and enthusiastic KIVFD recruiter.

Once the decision was made to establish their own fire department, the Kent Islanders gained momentum. On February 14, 1947, a group of men and women gathered at the Stevensville High School to organize and discuss the needs and obligations of such an endeavor. Mr. Curtis Larrimore of Easton, the senior instructor of firemen at the University of Maryland, and Mr. E.O. Willis of Centreville, president of Centreville's Goodwill Fire Company and vice-president of the Kent–Queen Anne's Fireman's Association, addressed the group. Costs were a preeminent concern.

According to *A Brief History of the Kent Island Volunteer Fire Department*, found in the cornerstone of the original firehouse, it was at this Valentine's Day meeting that Lee Bell was chosen to act as president until a first annual election could be scheduled. Benton Gardner was named secretary, and Barnes Legg, treasurer. The charter committee consisted of Henry Grollman, Edward Dawkins, Holton Harris, J.D. Sparks and George Taylor.

Other positions were filled on May 6, 1947. Ed Dawkins was elected the company's first chief. Henry Grollman's son Julius was named assistant fire

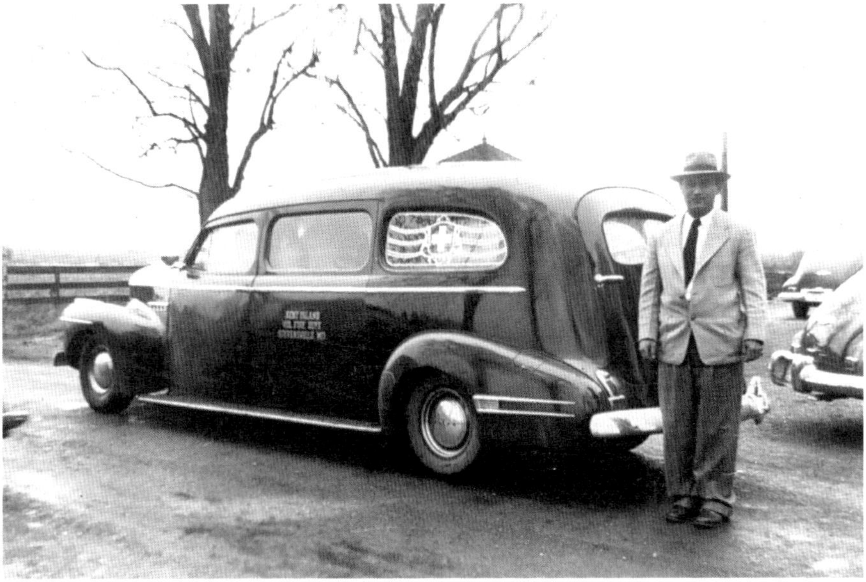

Sol Grollman, charter member and 1949 chief, with Queen Anne's County's first ambulance, a gift to KIVFD from the City of Baltimore. *Courtesy Mildred Haddaway.*

chief. Benton Gardner was made captain; Robert Snyder, lieutenant; and Julius's older brother, Solomon, sergeant. The superintendent of machinery was William E. Denny II.

The building committee was charged with finding a suitable location for the new firehouse. An unimproved Stevensville lot offered to the department in August was deemed too small. Dr. Theodore Cook soon donated a piece of land across from Stevensville High School.

At a December 7, 1947 meeting, six years to the day of the attack on Pearl Harbor, deliberations focused on firehouse specifications. All present were in favor of a 34- by 50-foot structure, as designed by builder Leonard Risley. Nine days later, the firehouse dimensions were expanded by 748 square feet, and two additional steel beams were ordered. On March 30, 1948, the deed for the lot was completed and submitted to the county by member J. Claude Lowery, and construction began.

The first ambulance call was for a car accident at Matapeake ferry. The *Queenstown News* reported on April 22, 1948:

> *The Kent Island Volunteer Fire Department's new ambulance had its first emergency run on Sunday evening carrying three patients to Easton*

> *Hospital from the scene of an automobile accident near Matapeake. The ambulance was driven by William Thomas, assisted by Julius Grollman and James White. Dr. Theodore Sattelmaier ordered the ambulance.*

The first fire call occurred later that year. The *Queen Anne's Record-Observer* of November 25 included a small story that a large brick farmhouse on the old Church Farm near Stevensville was almost completely gutted by fire, despite the presence of the Kent Island firemen.

Encouraged by his mother, a member of the ladies auxiliary, and Julius Grollman, Robert "Bobby" Norris was in high school when he joined the department in 1952:

> *When I was in the eleventh grade, our principal, James White, agreed to let the boys who belonged to the fire company leave class when the siren went off. We would run over one another heading for the firehouse. If you had a fire, you called either Denny's Garage or Grollman's Store. They'd set off the siren and then they would call the firehouse to tell us where the fire was.*

Babe Grollman was the youngest of Henry and Lena Grollman's children. "If you were asleep and Juls or Sol hit that button, the siren would go off and you'd shoot right straight up in bed."

Bobby Norris continues:

> *The first one there got to answer the pay phone inside the front door where they'd tell you where the fire was. As the members arrived, everyone was shouting as they hustled to get their gear on. The goal was to get on the first truck. Sometimes you had to run behind it and jump on as it pulled out; the intelligence and success of that depended a lot on who was the driver.*

Samuel "Kenny" Bullen—son of Josh and Sara Bullen, two of their era's KIVFD driving forces, and a life member with at least twenty years of service himself—says that a shortage of grown men available to respond created a situation where "most men worked on the water or farms. Sometimes high school kids were the only ones who were able to get there."

In Ray Baker's day, "We were the grunts. The leaders were businessmen, people who knew how to organize." As for why he joined the fire department, Ray says, "I joined right after graduating from high school, sitting in a car with Benton Gardner. We worked at Matapeake. He was encouraging people to join the fire department and suggested I join. He was a good salesman."

Curtis "Curtie" Chance was sixteen when he joined in 1966. Membership in the department seemed to come natural:

> *One of my earliest memories is when, in 1953 or 1954, Wendell Shawn's barns, where Kent Manor Inn is now, caught fire. It was a terrible thunderstorm. Lightning struck those barns and burned them down. My mother held me up in the air. There wasn't any Harbor View, no Catholic church and I could see that fire from Chester. I remember that to this day.*

Kenny Bullen joined in 1961:

> *Most charter members were still fairly active when I came in. Orem and Donald Thompson. George Taylor. My father. Mr. Billy Lane, Julius Grollman, Benton Gardner and Leonard Risley, they were all good men.*

Harry "Link" George has been a member since 1967. Link, a Department of Defense employee and KIVFD's longest-serving current active member, says:

> *I was twenty-four when I joined. I heard sirens going by my house and something triggered. I soon hurt my back playing softball, and Ray Baker*

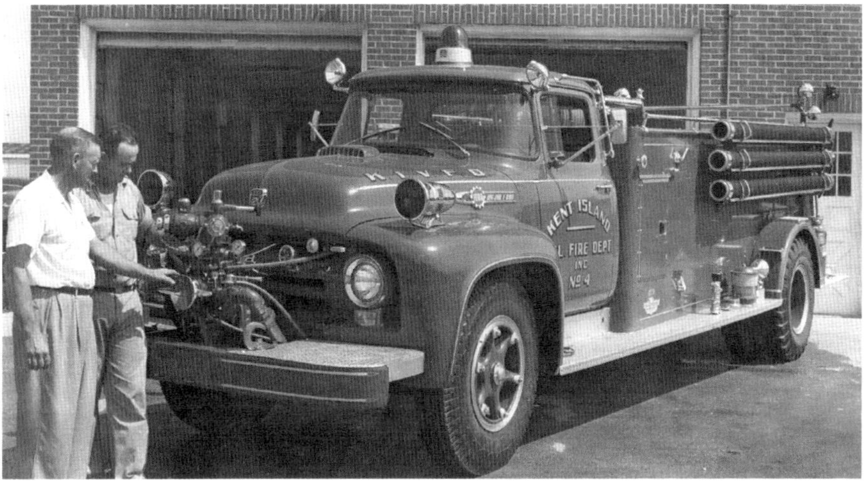

John Holden (chief, 1954–56, 1959–64) and Leonard Risley (president, 1954–56) inspect Engine #4. *Courtesy Joan Smith.*

took me in the ambulance. He said, "When you get feeling better, come on down the firehouse." The night I joined was the first time I ever went there. Everybody knew everybody. It's still friendly, but those guys were all hometown friends. From the minute I joined, I felt I'd been a part of it for twenty years. That very first time, we had a marsh fire at Love Point and I rode with Calvin [Duncan] *and Duke* [Moore], *sitting in the middle with my foot on the siren. Didn't even know it was the siren, until they told me to get my foot off it.*

Past chief and oldest active member Paul Fuld recalls:

I moved here in 1967. I was building my house and used to go to Tastee Freez for lunch. There was an accident at Castle Marina Road and Route 50. I'd been active in the Ferndale fire company in Anne Arundel County. When I saw the accident, I went over to offer my help. I ran into Duke Moore, who introduced himself. He encouraged me to join, and I've been here ever since.

Keith Baker, son of Ray, was a member from 1974 to 1983:

My father was in the fire company. All my friends were joining. I really got into it in 1971 and '72, before I was old enough to join. It was in my blood. My mother died in 1968, and my father lost interest. He quit, but it was pulling at me. When I got off school, the firehouse would be the first place I'd go.

Julius Grollman's son, Marc, feels the same way:

I joined soon as I turned sixteen. Dad, Uncle Sol and Uncle Alvin, Duke and Billy Moore were all in it. Keith Baker, Bobby Timms and Leon "Hammie" Collins, we all joined right about the same time. I'd hung around there all my life, but now I was a fireman.

Kent Island was in the middle of a growth spurt. After the first Bay Bridge span opened in 1952, predictions of an overnight onslaught of new construction proved premature. Speculators were a bit ahead of their time. Kent Island's real population explosion occurred between 1970 and 1980 at an astonishing rate of 113 percent. The bridge's three-lane second span opened in 1973. In the 1980s, sewer systems were installed on parts of the island to support and encourage the growth. New residences and businesses went up quicker than the natives were prepared for, despite having had

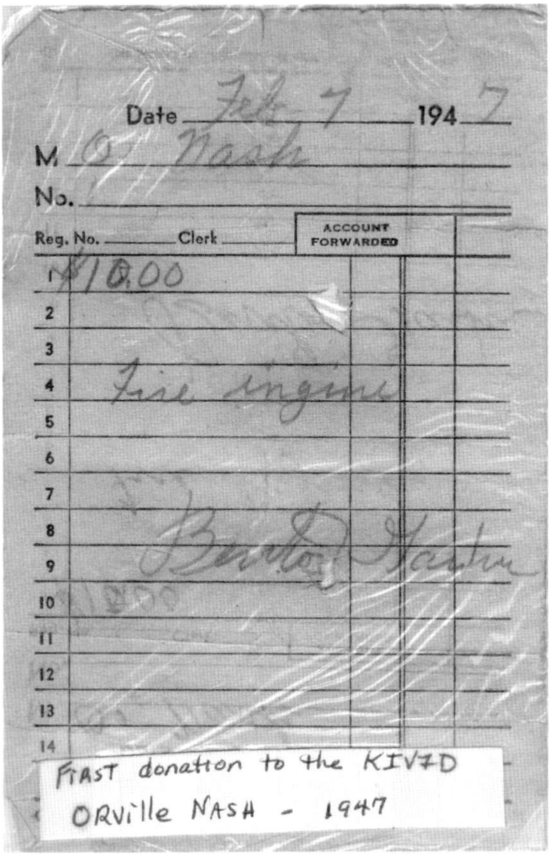

Captain Orville Nash of Chester officially donated KIVFD's first ten dollars to Benton "Crust" Gardner. *Courtesy Kent Island Heritage Society.*

decades to do so. In 1984, Queen Anne's County adopted its first building codes. Planning and Zoning issued a record-setting 1,214 building permits that year. Three-quarters of them were on Kent Island.

The *Queen Anne's Record-Observer* of April 4, 1975, noted:

> *For the first time in 23 years, Chestertown's volunteer firemen did not lead the list of "alarms answered" compiled by the Kent & Queen Anne Volunteer Fireman's Association. Kent Island Volunteer Fire Department responded to 93 fires and 10 assists in 1974—three more calls than Chestertown answered during the same period. Kent Island also led in ambulance calls, responding to 301 during the year.*

It was a trend that continued as Queen Anne's County became the fastest-growing county on the Eastern Shore.

Part II
ANGELS LIVING ORDINARY LIVES

I've seen and met angels wearing the disguise of ordinary people living ordinary lives.
—Tracy Chapman, American musician and recording artist

Volunteers and Other Characters

The life of a volunteer emergency responder is the same as everybody else's. But different.

Most volunteers have full-time jobs yet devote twenty to forty hours to the fire department every week. There are holiday celebrations, children's birthday parties and romantic dinners for two, but everyone involved knows that the volunteer may disappear at a moment's notice. They park their vehicles and prepare for bed in certain strategic ways. One minute they're doing commonplace household chores, attending church or sleeping, and the next their adrenaline shoots to measurably stressful levels. They're on their way to fight fire, rescue an injured child or recover the body of a lifelong friend, all at the risk of their own safety.

For free.

When that 1948 Mason jar time capsule was discovered in the original Kent Island firehouse in February 2008, one of the items inside was a pocket-sized Bible with an American flag on the cover published by the YMCA for members of the American Expeditionary Forces. The handwritten inscription inside reads: "James S. Ball, A.E.F., Mayen, Germany, April 6,

1919." Ball was a Stevensville High School teacher and artist who lived in the firehouse while assisting the department in the 1940s and '50s. One of Ball's legacies was the handmade and beautifully painted Nativity scene that still graces the firehouse at Christmas.

Former ambulance director Bobby Norris was a good friend of Jim Ball. "He was a gentleman and a talented artist," Bobby says. "He lived in the firehouse for several years, and I believe the department took responsibility for his funeral and headstone when he passed away."

The officers of the early Kent Island Volunteer Fire Department tended to be older, more established business and community leaders. The workforce consisted of younger men, some of them teenagers still in school.

Billy Denny contrasts the duties of each generation:

> *My dad was so involved he was gone a lot on firehouse business. He'd put me in charge of the garage. I never officially joined, but many, many times Juls Grollman would call and tell me he'd be by in two minutes. He'd slow down a little and I'd jump on. Sometimes it was so cold and I just got out of the shower and I'd freeze. I said, "From now on, I'm riding up front." He said, "I don't want to slow down that much."*

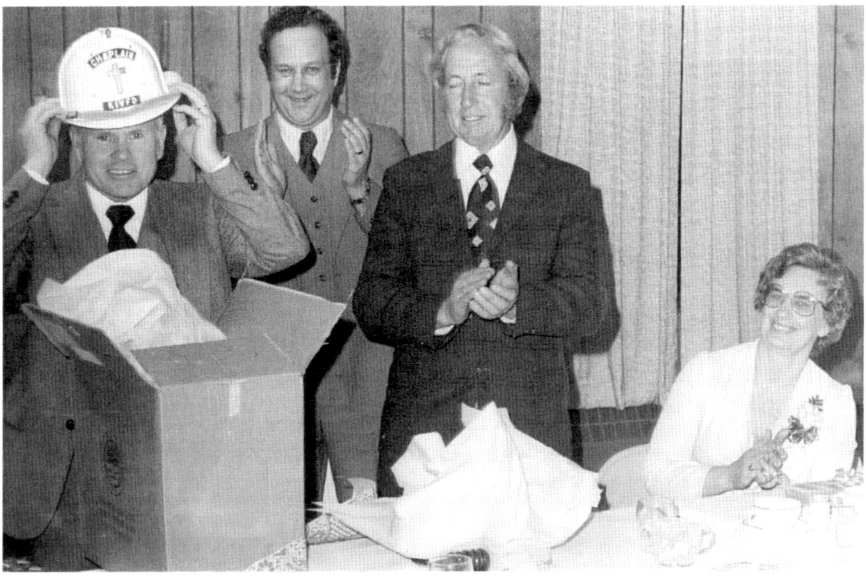

Honorary life chaplain Reverend Roy Phillips is presented his official helmet by Link George and then-president Warren Sadler, with Mary Lou (Fuld) Meadows. *Courtesy KIVFD.*

Chuckling, Billy praises his deceased friend. "Julius Grollman was as true to the fire company as he was the family business."

John Holden served as chief from 1954 to 1956 and again from 1959 to 1964. John was born in Queenstown in 1920 and worked as a guard at the Eastern Correctional Institution in Church Hill, Maryland. When John died in August 1970, popular local columnist Ralph Hoyt wrote:

> *He probably held every position of importance in his beloved Kent Island fire company. I'm sure that John Holden saw as much suffering and death as any other man in the county. It always seemed to upset him considerably, but he always answered the next call without hesitation.*

"He was a dedicated role model," acknowledges Bobby Norris.

Ray Baker laughs. "Only a few of us worked on the ambulance. John Holden would get excited and blink a lot when he drove. We'd be flying down the road and he'd get to blinking. He used to scare me to death."

Despite all the good-natured joking, one thing is acknowledged and agreed upon when Paul Fuld, a chief himself in 1978 and 1979, says, "John Holden was a good chief, the type of man you'd follow. He knew his job and was good at it."

Benton Gardner was a contemporary of John Holden and had been devoted to the KIVFD since the beginning. He served as chief in 1950, 1951, 1957 and 1958. Benton Gardner and John Holden, along with Ed Dawkins, Julius Grollman and Jul's brother Sol, together held the position of chief for almost two decades.

Department historian Dolores Timms points out some early unsung heroes:

> *Ray Baker and his father, Walter, wired the old firehouse for communication. He and his father were longtime members and very loyal. The Thompsons were very involved, very active. They were bricklayers. Orem Thompson built the firehouse.*

Babe Grollman says:

> *If they were building something and the fire siren went off, they'd drop what they were doing and take off. Of course, bricklayers were getting a dollar and a quarter an hour, carpenters 90 cents. If you left to fight a fire for a couple hours, you weren't losing much.*

Kent Island's hometown newspaper, the *Bay Times*, reported on December 25, 1964:

> *The Kent Island Fire Company is starting the New Year with a new chief. Last week the Island Firefighters voted in Dudley Moore of Stevensville to the top spot in the company. Moore replaces longtime chief, John Holden, who moves into the second spot as Moore's assistant.*

In her 1997 speech at the department's fiftieth-anniversary banquet, Dolores Timms asked, "Who can forget Duke [Moore, chief 1965–1977], with his string of ground prize tickets in hand to collect the first dollar you spent each year at the carnival or his many other deeds of service to the company?"

Kathryn "Kitty" Duncan (ladies auxiliary member since 1970) lays it out flat: "You couldn't tell the story of the fire department without Duke. Duke Moore was the fire department."

William Dudley Moore was born in 1931. He grew up in Centreville, where an eighth grade teacher nicknamed him "Duke." He claimed not to know why. In 1949, he started work as a veterinarian's assistant, a position he held for over twenty years. He married a Kent Island girl, Nellie Marie Clark (better known as Jackie), in 1953. Jackie came from a large, inseparable family and had graduated from Stevensville High School a year earlier. Duke moved to the island so his new wife could remain close to home. He joined the Kent Island Volunteer Fire Department in 1955. Jackie gave birth to their son, William Dudley, in 1956.

Calvin Duncan, chief from 1980 to 1991, says of his mentor, "Duke Moore is one of the greatest that's ever been in the fire department. It was his life. He was hardheaded, but there was never a more dedicated man."

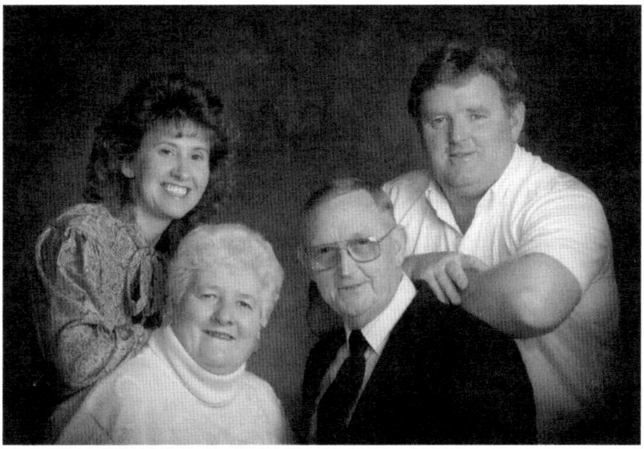

The Moore family: Joyce, Billy, Duke and Jackie. *Courtesy Joyce Moore.*

Tracy Schulz (chief, 1992–97, 2000–06) adds, "Duke treated the firehouse like it was his own, and he expected others to treat it with respect like it was his."

One of Duke's signature moves was to always be first to arrive at the firehouse. Members who knew him still wonder out loud how he got there as quick as he did. There's a popular suspicion that he slept in his clothes. Young Kenny Bullen lived only a couple houses away from the department, yet from the other end of town Duke would "be there when I got there." And he'd be fully dressed in a button-down shirt. "He didn't just throw on a T-shirt," remarks current president and member since 1978 Jody Schulz.

Third-generation volunteer Keith Baker spent a lot of time with Duke:

> *You had to know him to appreciate him. The tone would go off, and Duke's wife, Miss Jackie, would call to see if I was on my way. Most times I was. When I first joined, I slept in my socks and kept my pants straight down into my boots, ready to go. I'd meet Duke at the firehouse and we'd go on the call. I was quick, but generally when I'd get there Duke would be coming out in the ambulance or he'd have the door coming up and I'd go in under the door. One night I got to the firehouse and Duke was standing at the door. In his rush to get to the firehouse, he'd somehow put his pants on inside out and he couldn't get to the key in his pocket.*

After Dr. Franklin S. Wharton passed away, Duke helped the veterinarian's family on their farm for a year or so and then went to work on the Bay Bridge in 1970. He was a maintenance supervisor when he retired. Over the years, Duke and Jackie served the Kent Island Volunteer Fire Department with unequaled devotion. Their son, Billy, followed in their footsteps. Duke Moore passed away in 1993. Dolores Timms laments, "When Duke died, to not see him on the front of a fire truck, swinging his feet and waving at everybody who went by was a shame."

Kitty Duncan feels the same way. "He was an inspiration to everybody. I doubt there was a family on the island whose lives weren't touched by him. It would be impossible today to be as dedicated."

Duke ran thousands of calls over the years. It's no coincidence that the connecting road near the old firehouse between Main Street and Routes 50/301 is named Duke Street.

Billy Moore died unexpectedly on December 21, 2002, at the age of forty-six. Survived by his wife, Joyce, Billy was an avid golfer who also loved Eastern Shore hunting and fishing. He was faithful to the fire department all his life.

Jackie Moore passed away six months later. She was the last of her five sisters and four brothers, all committed KIVFD supporters.

Calvin Duncan joined the Kent Island Volunteer Fire Department in 1961. In 1965, a countywide alarm system was instituted. This alarm system assigned each of Queen Anne's County's eight existing fire departments a station number. Based on geography, south to north, KIVFD was designated Station 1. Trained by some of the company's original leaders, Calvin Duncan served as Station 1 chief from 1980 to 1991. Past president (1988–1999) Lou Johnson, a masonry contractor, recalls Calvin living across the road from the old firehouse. "He and Duke, when we had Engine 11, you knew when it went out they were on it."

Calvin was the popular bridge between the old guard and the newer generation of volunteers. Dolores Timms says, "The young guys really liked Calvin. Calvin had a way about him they related to. Calvin and Duke were alike in many ways."

Don Lewis, who as an overseas employee of the federal government served in the British Fire Services for five years, thinks Calvin was the "best chief I've ever known."

Others agree. Bruce Coursey belonged to the fire department for almost twenty years from the late 1960s. "Not to take anything from anybody else," says Bruce, "but Calvin was a great leader. I call him the five-foot giant."

Third-generation member Marc Grollman recalls:

> *Calvin never got upset like some of the older guys. He hung out with us more. I can't remember ever arguing with him. Right after he became chief there were a bunch of real bad house fires where there'd be nothing left but the foundation. We used to kid him and call him the "Foundation Saver." If somebody called the fire company because they smelled smoke in their house, when we got there Calvin would send me and Tracy* [Schulz] *in because of the size of our noses. He'd say, "Go in there and see if you smell anything."*

For his part, Calvin Duncan thinks back to one particular youngster who was fascinated by the fire department and matured into one of its most devoted members. Calvin reminisces, "I remember going down Cox Neck Road in the fire truck when Tracy Schulz was a kid, and he'd get behind you peddling the hell out of his bicycle to keep up. Man, you'd see him coming."

Life member Tracy Schulz is the youngest son of restaurateurs Oscar "Sonny" and Betty Schulz. His older brother, Jody, was in the fire department,

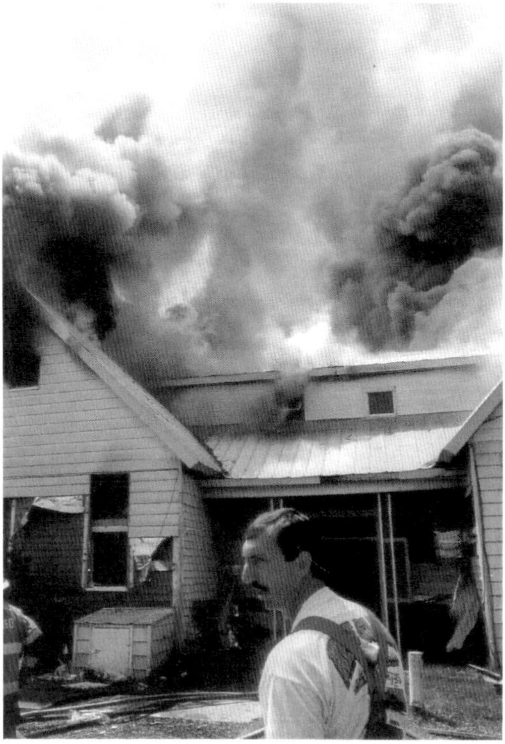

Tracy Schulz has answered an astronomical number of emergency calls and served in various leadership roles since joining KIVFD as a teenager. *Courtesy* Bay Times.

and Tracy joined as a fifteen-year-old junior member. Paramedic Josie Simons once said in a speech that Tracy's desire to ease the pain and suffering of others was just one of the reasons he's the "the kind of man you want your son to grow up to be." The Schulz home was broken into when Tracy was a kid, and a number of items were stolen, including the television. Tracy was infuriated that he couldn't find the emergency scanner, his electronic connection to the fire department. He relaxed once he saw that the monitor had merely been moved so the thieves could get to the TV.

Of course, as with any such organization, rookies were tested and taunted. When asked, Buddy Thomas, stalwart member since 1975, gets a laugh from his contemporaries with the reminder, "You had to be here three years before they called you by name. First three years your name was 'Stupid SOB.'"

When people deal with trauma and tragedy on a regular basis, one of the release valves is a fine-tuned sense of humor. Dolores Timms's oldest son, Robert "Bobby," a member from 1975 to 1993, has a number of favorite stories. This is one of them:

> *We were coming back from a call one night, and Billy Moore kept complaining his chest felt tight. We said what do you mean, Billy? Are you having chest pains? He said, "No, my gear just feels tight. I must be getting fat." Well, he'd jumped up in the middle of the night and threw something on to answer the call. When we got back, he started taking his bunker coat off, and he had on*

> *one of his wife's nightgowns. We laughed so hard at him, but that's the kind of stuff happens when you get up in the middle of the night to go put out fires.*

When discussing good times, the aforementioned Bruce Coursey is a name that's inevitably part of the conversation. Bruce grew up on the island and is physically described by most people as "tall and lanky." Despite a wild side, Bruce was an impassioned and effective volunteer. Calvin Duncan thinks of his onetime colleague this way: "Bruce Coursey was a real clown, but when something needed to be done, he was ready to do it. And soon as the situation was over, he was right back to being Bruce."

Curtie Chance says, "I used to jump in with Bruce on the way to a fire and I'd say, 'Don't go up this road like you're too crazy now.' We'd go up there sometimes on two wheels."

Bobby Timms chuckles, "Bruce didn't drive trucks, he aimed them."

Terry Smith is a well-known area radio personality. Her on-air name is Terry Alley. Her husband, Tucker, comes from a family devoted to the KIVFD. Terry's mom, Joyce Dennis, was a longtime ladies auxiliary member and past president before passing away in April 2010. Over recent years, Terry has been a driving force in the department's fundraising efforts. Discussing Bruce Coursey, she says, "When I was pregnant with our oldest son, I told Tucker, 'When I go into labor don't call the ambulance. I love Bruce but do not want him delivering my baby.'"

Second assistant chief Tracy Schulz recalls, "Bruce would come to the firehouse in his underwear and wear his bunker pants home."

Bobby Timms tells the following Bruce Coursey story:

> *A guy drowned one night at Stevensville Bridge. Keith Baker, Marc Grollman, Hammie Collins, Mark Harrison and a bunch of them went to the car show in Baltimore. They called wanting to know what happened. We told them a man drowned, and we were at the firehouse waiting for the undertaker to come get him. Bruce got on the stretcher. We sprayed his shoes with water, covered him up, made the sheets wet. Here comes Keith Baker. He had few drinks in him. I asked him if he wanted to see the victim. Man, Keith and Hammie and a couple of the others crawled into that ambulance. When they pulled that sheet back, Bruce sat up and roared. Those boys almost burst the roof off trying to get out of there. We laughed so hard we were crying.*

Bobby's mom, Dolores, thinks of Bruce fondly. "Bruce would tell it like it was even when you didn't want to hear it. He loved to joke around; he'd

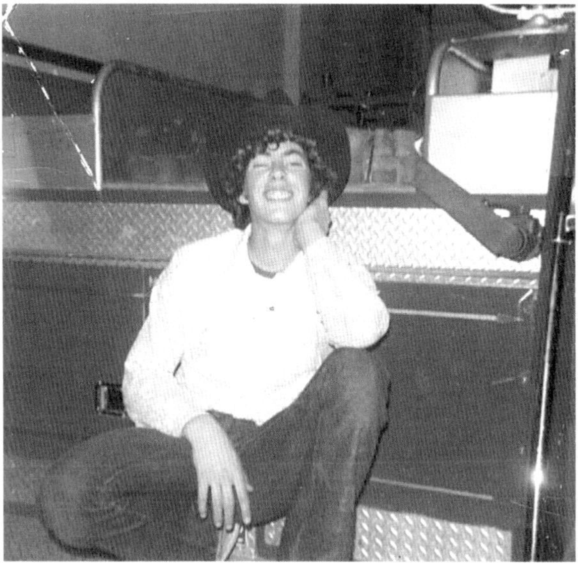

Brucc Coursey, a devoted volunteer and an unforgettable character, at age eighteen. *Courtesy KIVFD.*

do anything—the crazier the better." She adds emphatically, "And he was a great fireman. We miss him since he moved away."

Kathy Hemler was one of Kent Island Volunteer Fire Department's top Emergency Medical Services responders before becoming its first female firefighter.

The Hemler family—Joe, Evelyn, Debbie and Kathy—moved to Kent Island in 1975. Purchasing and reopening what was long known as Frampton's Store in Stevensville, the Hemlers quickly endeared themselves to the locals. Joe and Kathy joined the KIVFD ambulance crew. Evelyn and Debbie became members of the ladies auxiliary.

Kathy soon decided to expand her skills and diversify her service. Her introduction to firefighter politics was not an easy one. Bobby Timms remembers the controversy a woman brought to a traditionally male force:

> *That was a tough time. Those old guys, and some of the younger ones, were hard on Kathy. They weren't prepared for women who were so outwardly strong-willed. The company was split. Some thought it was a good thing; others felt things were changing too much. There was a lot of controversy. Those were dark days for the fire company.*

Former treasurer and vice-president Keith Baker saw how trailblazer Kathy Hemler proved herself despite such obstacles. "I give Kathy a lot

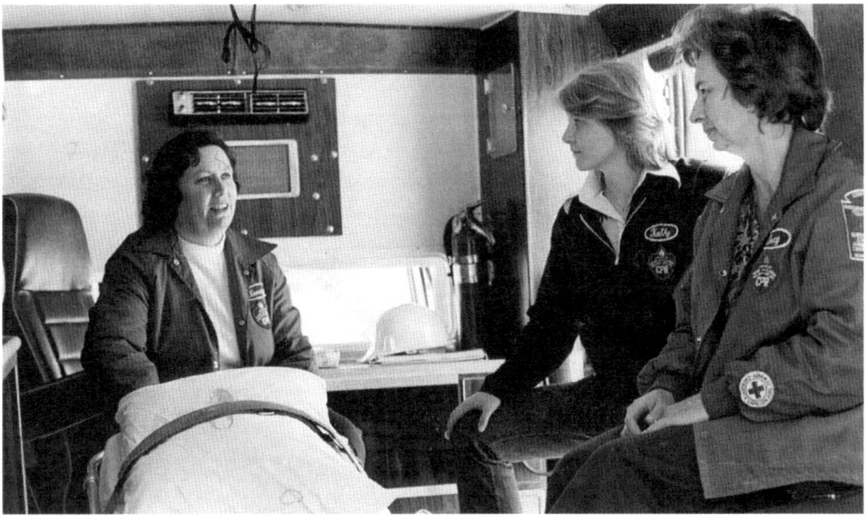

In 1978, Nancy Sadler, Kathy Hemler and Mary Lou (Fuld) Meadows were among the vanguard of Station 1 women to provide emergency services. *Courtesy KIVFD.*

of credit. I saw her outdo male firefighters." He considers the bottom line. "The question was always if I took her into a fire scene with me would she watch my back, and I believed she would."

Kathy Hemler later helped establish the county's Advanced Life Support system and broke more gender barriers when she joined the Maryland State Police. She says she knew joining the masculine worlds of firefighting and police work would be no easy task, and she holds no ill will toward those who resisted her inclusion. "It wasn't like I didn't expect a challenge," she says with a smile. The Hemlers moved and are no longer active in the KIVFD.

"And then," Don Lewis (president, 2002–04) says,

> *Grace Gardner came along. Like Kathy she mostly worked on the ambulance, and she wasn't taking any crap. She's still a valued member, the current EMS treasurer, and she's tough. I used to see her at the old firehouse running across that apron. She'd be getting dressed with her bra in her hand. She'd get in that ambulance, put her bra on and go.*

Today, KIVFD has a handful of female members, mostly on the ambulance crew but also a couple of firefighters. Their presence is expected, part of the professional dynamic of a modern emergency response organization.

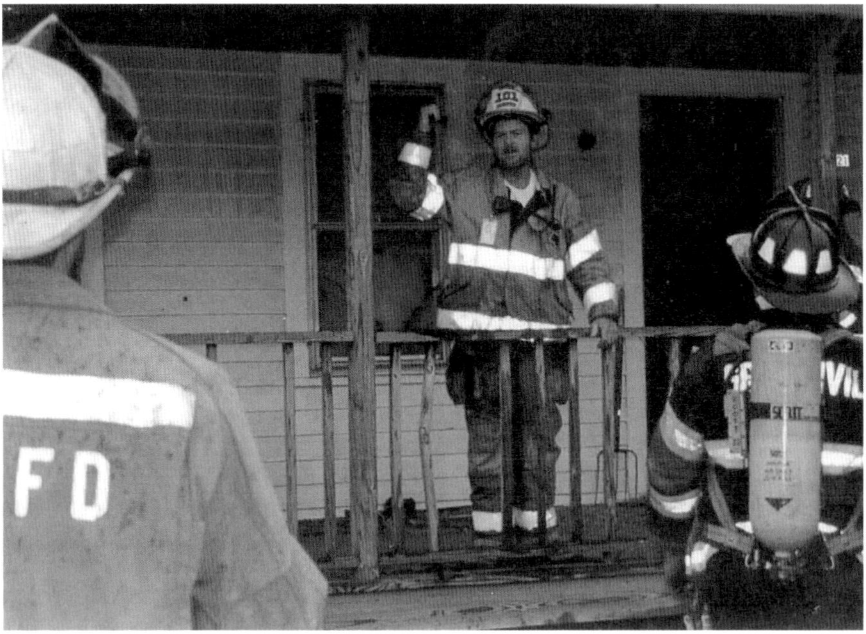

Dougie Thomas (1961–2003) was a dedicated KIVFD member and a Kent Islander through and through. *Courtesy Thomas family.*

Native Kent Islander William Douglas "Dougie" Thomas was born on October 21, 1961. His grandfather was a founding member, and with Alfred Thomas's encouragement, Doug joined as soon as age requirements allowed. For over twenty-five years, Dougie Thomas was a leading responder and held numerous elected leadership positions, including chief from 1998 to 2000 and a seat on the department's board of directors.

On the Sunday morning after the terrorist attacks of September 11, 2001, Dougie was responding to a reported boat fire at the Kent Island Yacht Club at Kent Narrows when the fire engine he was driving struck a utility pole and overturned. He suffered serious spinal injuries.

Dougie Thomas was a stubborn man. He defied expectations and struggled to hold on to life for almost two years. Unfortunately, a July 2003 issue of the *Bay Times* announced:

> *One day after Queen Anne's County dedicated its new memorial to fallen heroes, the county lost another. He was William "Dougie" Thomas, second assistant chief of the Kent Island Fire Department, who died July 9 at the University of Maryland Shock Trauma Center in Baltimore as a result of*

injuries sustained while responding to a fire call in September 2001. He was 41. This week, family, friends and fellow firefighters took part in an emotional celebration of his colorful life, and then bore his casket atop a fire truck to the Stevensville cemetery, where he was laid to rest with full fire department honors.

Dougie left behind his beloved wife, Renee; his parents; his children; and a grandchild. He's the only member of the Kent Island Volunteer Fire Department to sustain life-ending injuries in performance of his duties.

That August, the paper offered better news: the induction of Link George and Paul Fuld into two separate firemen's halls of fame.

Link George has been an active fire department member for more than forty years. Introduced to the KIVFD by Ray Baker, Link has served in a number of official positions, including president and chairman of the board of directors. When inducted into the Fireman's Historical Foundation of Delmarva's Hall of Fame, then-president Don Lewis described Link's service. "Link went to work doing what Link knew best, helping his community and being a good fireman."

In a later interview, Link said, "I wouldn't have been here this long if it weren't for the camaraderie." He grinned, "And it sure hasn't all been pretty."

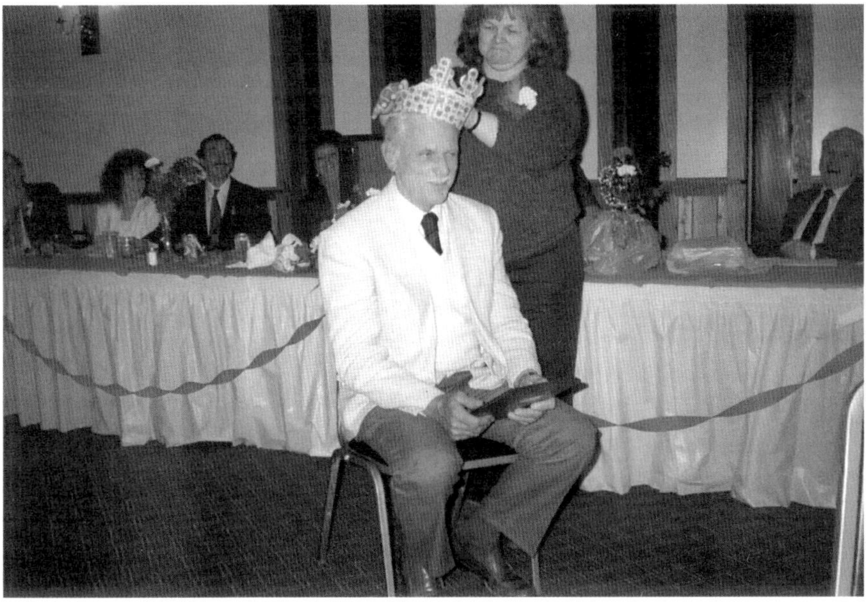

February 1993. Paul Fuld grimaces as he receives the King of the Bingo Hall Award from Grace Gardner. *Courtesy Kent Island Heritage Society.*

Lou Johnson joined in 1986 at the urging of his son, Steve, who was already a member. Lou compares Link George and Paul Fuld. "Link's been here the longest, but Paul's our oldest active member."

Paul Fuld was inducted into the Eastern Shore Volunteer Fireman's Association Hall of Fame. Paul joined the department soon after moving to the island and has served in various leadership positions, including a stint as chief in 1978 and 1979. President Jody Schulz noted that "back in the day, Paul and his ex-wife, Mary Lou, were instrumental in getting the ambulance out. Those two plus longtime treasurer Clinton Phillips and [past president] Warren Sadler's wife, Nancy, were extremely active in those regards."

Ladies auxiliary member and current secretary Shirley (Baxter) Thomas says, "A lot of laughs have been had at Paul Fuld's expense. He's fun to mess with." She continues, "And everybody loves him to death. Everybody calls Paul 'Friend.'"

"It's my trademark," Paul Fuld says. He downplays his long service with a shrug. "I still do the same things I did twenty years ago. Drive, work wherever I have to—carnivals, fire conventions—and I run my mouth." He does, however, admit to maybe not being as fast as he once was. "Only thing different is now if I respond to a call, by the time I get out, the call's been cancelled."

Current chief Paul Schlotterbeck works as a fire investigator for the state, and in 1993 he was assigned to the upper Eastern Shore. After moving his family to Kent Island, Paul joined the KIVFD, and his knowledge, experience and dedication allowed him to move up quickly through the ranks. As chief, he's responsible for operations of the fire department. He thinks of his volunteer job as a professional obligation.

> *From the middle of the Bay Bridge to the Kent Narrows Bridge, from about Batts Neck* [since 1972, south of Batts Neck has been serviced by Station 9, United Communities Volunteer Fire Department] *to Love Point—I'm responsible for the safety of everybody who lives there.*

Service to the community runs in the family. His wife, Sherry, is an EMS lieutenant; his oldest daughter, Hollyann, an emergency medical technician and firefighter; and his youngest, Amanda, a cadet. The cadet program is available for junior members ages fourteen to sixteen and provides teenagers a safe introduction to emergency response.

Members such as Link George, Paul Fuld and Paul Schlotterbeck maintain a tradition of service that is the hallmark of the Kent Island Volunteer Fire Department, yet it's impossible for former firefighter and EMT Marc Grollman not to praise the volunteers of the past.

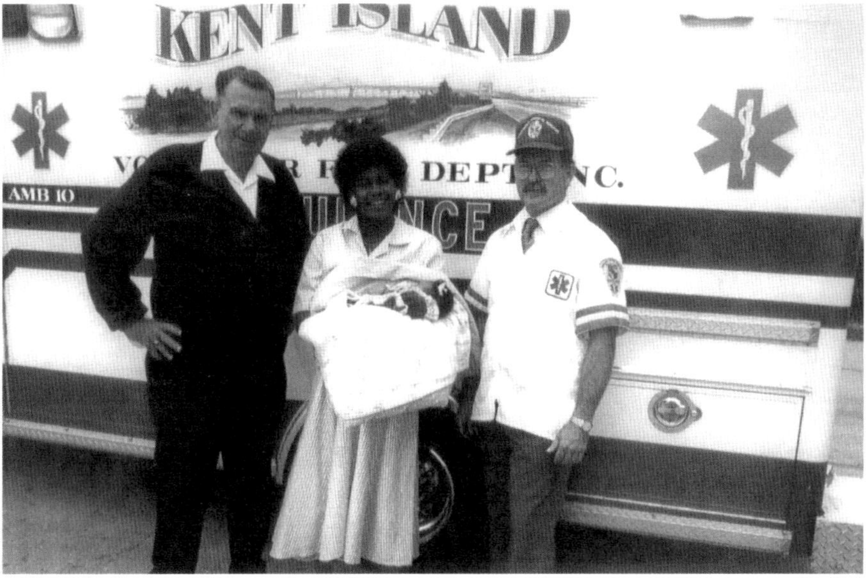

Delivering babies? Part of the job. The annual Jim Beavers Award, honoring an outstanding EMS volunteer, is named for the gentleman at left. *Courtesy* Bay Times.

"Duke and Jackie and Billy died. Those three were the heart of the fire company. But there were others, too. Jackie's sisters, all the Colliers—Louis, David, Hazel, Harmon, Brother—the Thompsons, the Lowerys." Marc is unembarrassed by his close-to-the-surface sentimentality. "Those people worked hard for their community. We should never forget them."

In their loyalty to the organization and the everyday selflessness with which they reach out to their neighbors, members of today's Kent Island Volunteer Fire Department honor the memory of all who came before. It's an inheritance they'll pass on.

The Ladies Auxiliary

Bobby Norris was a member of the Kent Island Volunteer Fire Department from 1952 to 1960. In a letter commemorating the department's fiftieth anniversary, Bobby wrote:

> *If the men were the blood and guts of the company, the ladies auxiliary was the backbone. The men were the brick and mortar; the ladies were*

> *the foundation. They gave the company a heart and soul in everything they did to support the fireman. It would be impossible to measure how much money they raised, how many meals they prepared and how many bingo games they organized over the years. Their individual and combined service, dedication and contribution were enormous. Many of them were married to the men members, and those strong family ties, that were so very important at that time, were a key element in the constant strength and growth of the company.*

In her speech honoring the same anniversary, Dolores Timms asked:

> *How can we forget the dedication of our wonderfully energetic ladies auxiliary who has been there for fifty years to aid the firemen? While the calls took the men out for long periods, the ladies were there to serve sandwiches and drinks. They have held bake sales, dances, talent shows, raffles, bingo, and election day dinners. There's no way to count the hours of carnival work they have done. And cakes? The ladies would say we've "baked a million of them."*

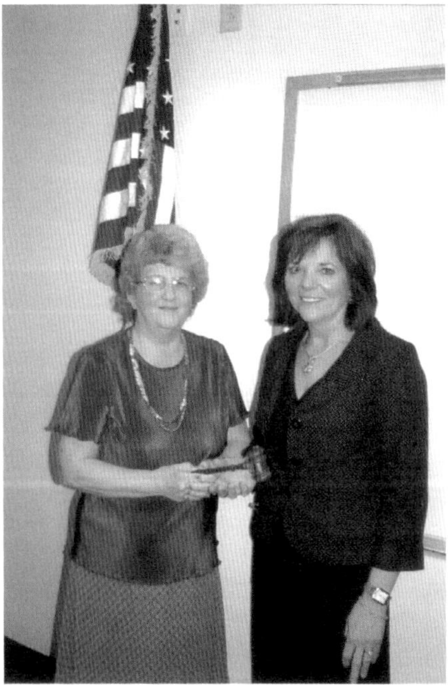

In 2009, Dolores Timms passed the presidency of the KIVFD Ladies Auxiliary to Shirley Sperl. *Courtesy* Bay Times.

However, past president Link George (1970–73, 1977 and 1986–87) speaks for most who know her when he asserts, "Dolores Timms is the heart of our ladies auxiliary."

Dolores was raised among firefighters. "Growing up in Queenstown, I learned when the siren blew, you got out of the way and let everybody through. My mom cooled her cakes on the stair steps, and my brother Milton never failed to step in one on his way down."

The male and female volunteers from Queenstown were supportive of the island upstarts. According to *A Brief History of the Kent Island Volunteer Fire Department*, on February 25, 1947, a motion was made to invite Mrs. John Cole and Mrs.

Harry Reynolds of Queenstown to come and familiarize the islanders with the function of a ladies auxiliary unit.

The following is from the first minutes of the ladies auxiliary of the Kent Island Volunteer Fire Department:

> *The first meeting of the Ladies Auxiliary was held at the Stevensville High School at 8 p.m. on March 25, 1947. Mrs. Emily Denny was acting chairman; Mrs. Virginia Bell was acting secretary. It was moved and seconded that the dues of the Ladies Auxiliary would be $.50 per year.*

At a meeting on April 8, Emily Denny, a popular teacher and local historian, was elected the ladies' first president. There were four vice-presidents, "one from each section of the Island." Miss Catherine Kirwan was elected secretary, and Mrs. Charlotte Severa, treasurer.

According to *A Brief History of the Kent Island Volunteer Fire Department*, "The first venture of the auxiliary was a supper on April 11, 1947 from which $501 was cleared." That same document acknowledged that $950 in profits were made on "the Matapeake venture," and by 1948, there were over one hundred members of the ladies auxiliary.

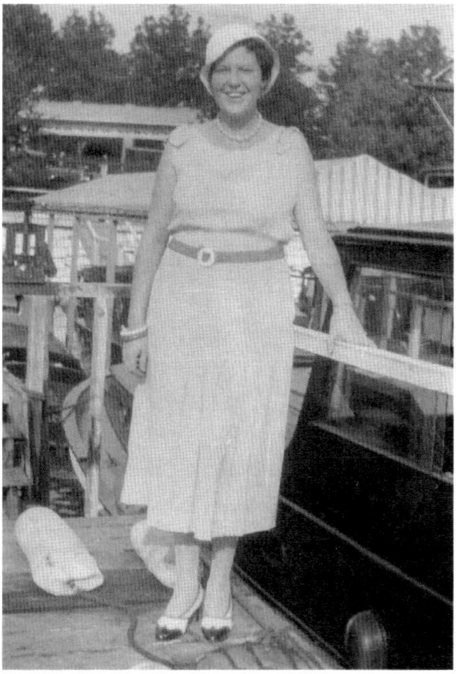

Catherine Kirwan's dedication and generosity to the Kent Island Volunteer Fire Department was enthusiastic and enduring. *Courtesy Kirwan Collection.*

Through her research, Dolores Timms learned of the original fundraising efforts. She says the auxiliary sent thank-you notes to Madeline Lowery and Hazel Collier for their cooperation in working that first supper, and the "Matapeake venture" referred to fire department and auxiliary members selling sandwiches and drinks to motorists waiting to board the Matapeake ferry.

Billy Denny was there. He says, "We carried a bucket of soft drinks in one hand and a bucket of sandwiches in the other."

The ladies auxiliary soon developed a widespread reputation for their real Eastern Shore–style home cooking. A July 21, 1948 letter from KIVFD president Lee Bell to a Baltimore group read:

> *The Ladies Auxiliary voted on your proposed visit and informs me they will be happy to serve you on Saturday evening August 14th. The Ladies will serve the following menu—Fried Chicken, Crab Cakes, Creamed Potatoes and Chicken Gravy, Lettuce and Tomato Salad, Green Lima Beans, Homemade Bread and Butter, Homemade Pie and Iced Tea. The supper will be served Country Style. Price—$2.25 per person.*

Billy Denny continues:

> *The ladies didn't get enough credit for how hard they worked. Ladies like my mother* [Ida Gibson Denny], *my aunt Emily Roe Denny and Miss Audrey Sparks. All those Smith girls, the Colliers and the Clarks worked very, very hard for a long time.*

Bobby Norris mentions a few others. "There was Lottie Taylor, Miss Dorothy Risley and George Grime's mother, Miss Beulah." He recalls them warmly. "I can smell their cooking in that kitchen right now. Those ladies were saints."

The KIVFD was a tightknit group. For Kitty Duncan, "It was a family thing. My dad and mom [Charles Lindbergh "Babe" and Reba Clark Collier] and all my aunts were members."

In those days, the ladies auxiliary went to the firehouse to make food on every call. Shirley Thomas, ladies auxiliary president from 1996 to 2000, says, "When the men woke up to go to the fire, we woke up to go to the firehouse. We'd be cooking breakfast when they came back."

Dolores Timms adds, "Our children always went to the firehouse with us when we worked. We were from the day you just took your children with you."

Cindy (Smith) Taylor, a second-generation ladies auxiliary member, holds fond memories of her mom's era of female volunteers:

> *Jackie Moore and Miss Reba Collier were a lot of fun. Whenever you went somewhere with Jackie, the party was always going to be in her room. She'd invite everybody over to her room for this party, you'd get to her room, knock on the door and she'd be sound asleep. I enjoyed when we'd go to Chestertown and Cambridge to visit hospitals. We'd buy little necessities*

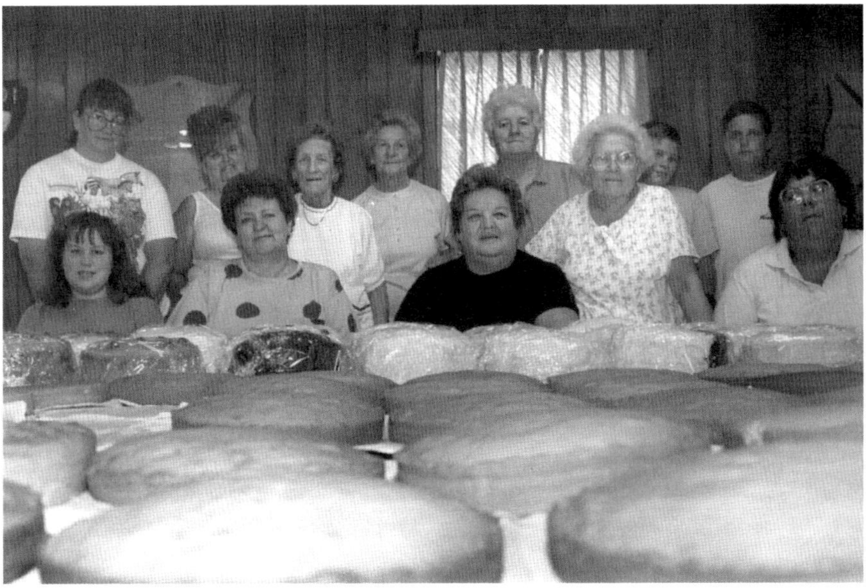

The ladies auxiliary has raised funds, supported its firefighters, threw memorable parties and baked an uncountable number of cakes. *Courtesy Kent Island Heritage Society.*

> *and gifts, get on a bus and put on a bingo for patients. It was so rewarding. It meant so much to them.*

Shirley Thomas was on those goodwill bus trips to area hospitals. "Tommy Pauls would drive us and we'd stop at Holly's restaurant on the way back. They stayed open all night long. It was so much fun."

Sara Bullen was one of that era's auxiliary leaders. Married to Josh Bullen, Sara was a link between the early membership and the next generation. Dolores Timms says, "When I went to the old, empty firehouse the other day, I could see Sara Bullen standing at the end of that counter. That was Sara's spot. Nobody touched Sara's spot."

"You'd better stay out of it," warns Kitty Duncan.

Dolores concedes, "She had that corner and she'd be icing those crushed pineapple cakes to perfection. I can see her standing there icing those carnival cakes."

Jody Schulz was not much more than a kid, but he knew "you didn't want to sneak up there and try to get a piece of cake. She wasn't having it."

Marc Grollman smiles. "I love food, and besides my mother's and my family's cooking, the ladies auxiliary was tops. Installation dinners, dances,

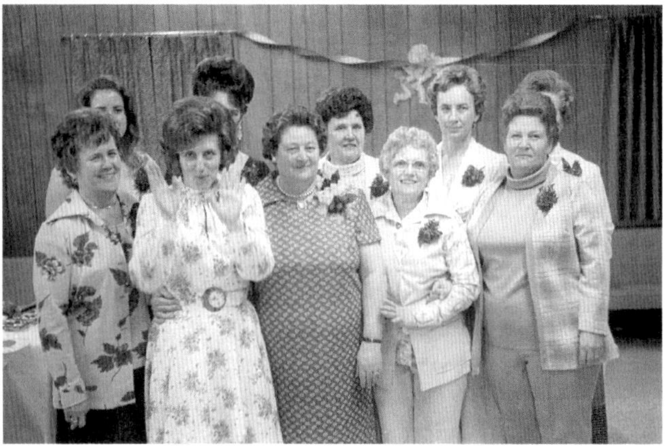

Janet Mann, Margaret Phillips, Sara Bullen, Mary Lou (Fuld) Meadows, Hazel Collier, Ruth Legg, Dolores Timms, Magdalene Walters, Claudia Angelini and Jane Legg. *Courtesy KIVFD.*

election days, controlled burnings—the ladies' cooking was always one of the best parts."

"There was nothing like Miss Jackie's rolls and Miss Sara Bullen's oyster fritters," declares Link George.

Cindy Taylor is quick to add, "You didn't mess with Miss Sara's oyster fritters, either."

Shirley Thomas shares the following Miss Sara story:

> *We were upstairs at the old firehouse playing bingo. If two players called bingo at the same time, one would get the money and the other would get a can of milk or a bag of sugar or something. I sat there all day and when I finally yelled bingo, Miss Sara got bingo at the same time. I took the can of milk and let her take the money. I said, "I'm not crossing her!"*

"She was one of the hardest workers and had a heart of gold," says Dolores Timms. "If you didn't know her and met her you'd think she was so mean, but she really wasn't. She had a way she wanted things done and that's the way they were done."

Dolores honors her predecessors:

> *The whole Clark family belonged, all the sisters. Jackie Moore, Madeline Lowery, Bunky Boone, Hazel Collier, Reba Collier and Elsie Smith. They did so much hard work and accomplished so many wonderful things. They really inspired me.*

"They worked until they couldn't work anymore," states Shirley Thomas.

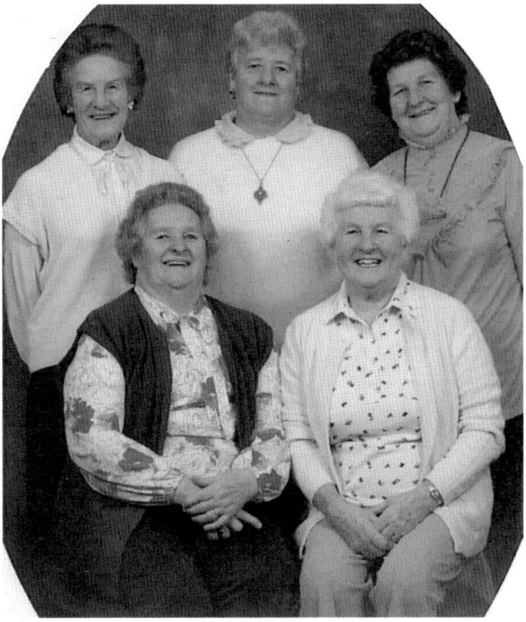

The Clark sisters: Hazel Collier, Nellie "Jackie" Moore, Mildred "Bunky" Boone, Reba Collier and Madeline Lowery. Elsie Smith is not pictured. *Courtesy Joyce Moore.*

Shirley Thomas notes that "Claudia Angelini was a very active leader for her time, and her family has been very involved." Claudia Angelini was ladies auxiliary president in 1978 and from 1981 to 1983. She died at the age of eighty-six in 1999. Claudia's son, Bill, now a retired police sergeant, joined the KIVFD as a teenager in 1966 and still helps out when and where he can. Of his mom's commitment to the department, Bill says, "There wasn't anything she wouldn't do. They could call at 3:00 a.m. and she was ready to help however she could."

At Christmastime, the ladies assist the men with Santa Claus duty. Shirley remembers:

> *As a kid, you'd go to the firehouse to see Santa and get a toy. There'd be a table with little gifts for your age group. Coloring books, puzzles, stuff like that. You'd pick out something. We'd be lined up around the table and down the stairs.*

"Now," says Kitty Duncan, "the fire company goes out into the community with Santa on a fire truck and gives out candy canes. Kids love it, and so do a lot of senior citizens."

Dolores Timms expounds on the ladies auxiliary's current outreach programs:

> *The last couple Christmases we took candy and gifts and toiletries to seniors in assisted living facilities. The residents really enjoy and appreciate it. We also have our Angel Child Program. The very first bingo game every week is the Angel Child game. Half the money goes to the winner; half goes into*

> *a fund for when we come across children, mostly cancer victims, who are struggling financially. These funds help with costs like traveling back and forth to hospitals or special trips and requests. We've helped ten or twelve children over the years. Some of them have passed, some have survived. The program is active when we have a child that needs us. We've helped people who were temporarily in a bind and came back later to make a substantial donation.*

Unfortunately, modern membership in the ladies auxiliary has waned. In 1947, there were fewer than thirty members of the fire department and over one hundred in the auxiliary. Today, the numbers have virtually flipped. One of the reasons for the change is that contemporary female volunteers might have their sights set on other service. Cindy Taylor explains, "Back then, women were strictly in the auxiliary. They weren't members of the fire department. Now it seems the ladies auxiliary has trouble getting new members because they want to be firefighters."

Dolores Timms has witnessed the changes:

> *All the women joined the ladies auxiliary. First off, they were all very community minded. When I got married, I started playing bingo at the Clark girls' house. While I was there, they started discussing the auxiliary and why didn't I join. I started going to different functions, and in 1968, I officially joined. It was a different time. Everybody was involved in the church and the fire department. In 1947, the ladies auxiliary had 108 members; now we have 28. Eighty percent of our members are over sixty-five. Since we've moved to the new firehouse, we've gotten five or six new members, but we have no really young members.*

Yet the flame lit in 1947 does still burn. In the minutes of a 1947 meeting, secretary Catherine Kirwan wrote:

> *We started as a group of women interested in the Fire Company. We were interested in it as a betterment for the safety and welfare of our community or we would not have done and worked as we did. Our Fire Company has organized Kent Island such as no other organization has been able to accomplish.*

Today, in addition to its other altruistic activities, the KIVFD Ladies Auxiliary can be counted on to donate between $10,000 and $15,000 to the Kent Island Volunteer Fire Department annually.

Part III
FIRE TRUCKS AND NO MAN'S GHOST

I can think of no more stirring symbol of man's humanity to man than a fire truck.
—Kurt Vonnegut Jr., author

Let no man's ghost return to say his training let him down.
—firefighter's adage

Apparatus and Equipment

While the Kent Island Volunteer Fire Department's 1947 building committee concentrated on obtaining a suitable site and the proposed construction of the firehouse, others worked to acquire equipment and apparatus. According to *A Brief History of the Kent Island Volunteer Fire Department*, the document chronicling the fire company's first year, on March 25, 1947, the department voted to purchase an International chassis from Mr. William E. Denny and a K-7 truck tank from the Oren Company.

Efforts to procure an ambulance originated from a different source. Mary Catherine Kirwan was the only daughter of Lemuel and Maude Golt Kirwan and granddaughter of state senator James E. Kirwan (1848–1938). Her grandfather served as a bay-faring policeman in Maryland's Oyster Navy, as an elected official and on various boards and committees. He established a general store and lumberyard on Dominion Road that was a center of Kent Island commerce and social activity for decades. Senator Kirwan helped lead the successful 1917 resistance against the

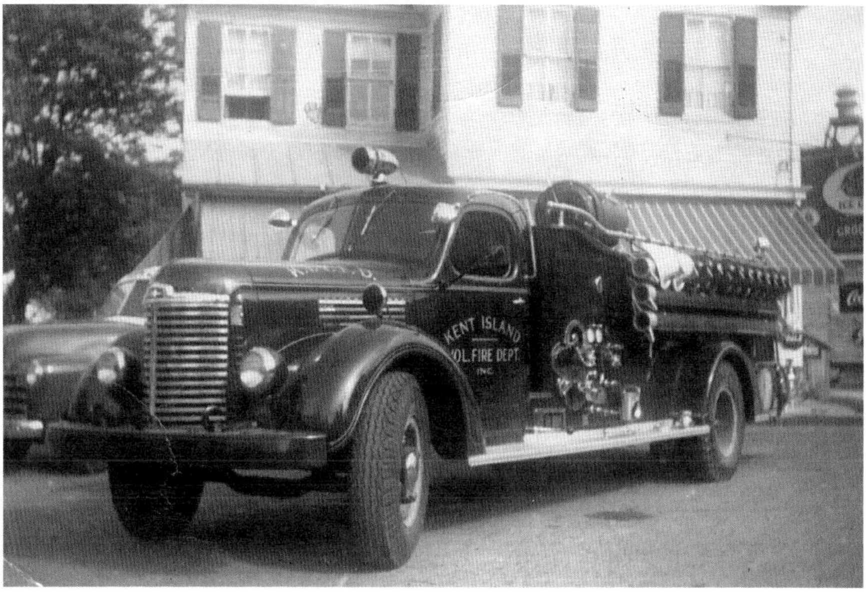

KIVFD's first fire truck was this 1947 International. Frampton's Store, a Stevensville landmark for decades, and the fire siren are in background. *Courtesy Joyce Moore.*

federal government's efforts to turn Kent Island into a weapons-testing ground. Thanks to a donation of the property, home and store a few years after Mary Catherine's death in 1994, The Kent Island Heritage Society opened and operates the James E. Kirwan Museum, a re-creation of a turn-of-the-twentieth-century country store.

The senator's granddaughter was born on April 18, 1913, and grew up to be a determined and energetic civic booster. Miss Catherine used her family's political connections to obtain a used ambulance from Baltimore mayor Thomas J. D'Alesandro Jr., father of the first female speaker of the U.S. House of Representatives, Nancy Pelosi. The gift was encouraging to Kent Island's newest civic organization.

The *Queenstown News* announced the ambulance's celebrated arrival via ferryboat:

> *Sunday, April 11, 1948, will be remembered for many years to come by all the members of the Kent Island Volunteer Fire Department. That day marked the "blessed event" of their first piece of Fire Department equipment. Amid cheering and horn blowing the first lusty wail of the siren was heard as it emerged from the Matapeake Terminal.*

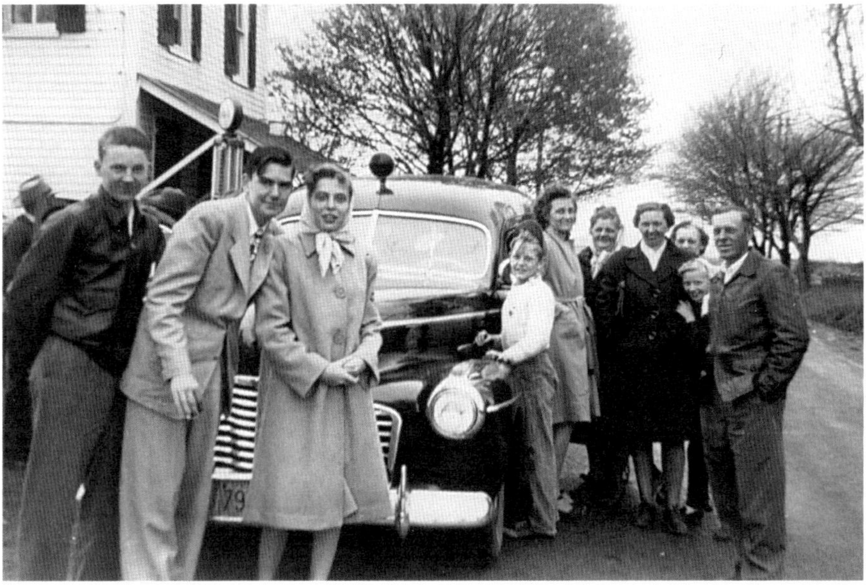

Coming to Kent Island aboard the Matapeake ferry in April 1948, the donated ambulance was admired by islanders of all ages. *Courtesy Mildred Haddaway.*

Local author and historian Nick Hoxter says, "Everybody was so excited when Solly Grollman picked that ambulance up. It was a big day."

On April 20, the ladies auxiliary turned over the ambulance's title to the fire company's equipment committee. On May 11, the rules and organization of the ambulance corps were put into effect, including a rate of fifteen cents per mile.

Founding member Billy Lane remembers, "We got the original ambulance before there was a firehouse. I believe we kept it in Grollman's garage."

Billy Denny, son of one of those founding volunteers, recalls:

> *I think we kept the ambulance in my dad's garage a while. It was a 1941 Buick. It was a nice ambulance and didn't cost us a dime. Every time I turned around, Juls wanted me to polish the ambulance. So we'd shine it until we almost took the paint off. One time we were shining that ambulance, and there was a big back door on it. I was standing on the inside wiping the outside and the wind came up. Blew that door to, and it caught me right across my stomach. It knocked the air out of me. Juls stood there laughing, and I said, "You bugger, I could have been killed!" He said, "Well, we've got an ambulance right here."*

By October 1948, the new firehouse was almost complete and awaiting service. The progress of the Kent Islanders in establishing their own fire department was big news. The *Baltimore Sun*, dateline October 15, reported:

> *Almost simultaneously, lights were turned on in the newly completed $25,000 Kent Island Volunteer Fire Department firehouse as a new $11,000 pumper rolled up to the door. Chief Edward Dawkins said the new engine was an Oren 1,000 gallon on an International chassis with 1300 feet of fire hose and 250 feet of booster hose.*

The same day, the *Queen Anne's Record-Observer* told readers:

> *Greeted by several hundred firemen, auxiliary members and just plain citizens, the shiny new fire truck rolled into Stevensville Friday about 9:30* [p.m.], *after several hours delay, but, according to all reports, the wait was well worthwhile and the people stayed late in the night to inspect the truck.*

Past president and retired state tax division employee Ray Baker comments on other early apparatus:

> *We bought most of our truck chassis from the Dennys, and we fitted them with equipment. They performed beautifully, hummed. First was that K-7 International. It had a powerful engine and served us for years. That International took forever to get going. Big old Number 11 was something to drive. Later, we bought a smaller truck because we needed something to get around in the fields and woods and had a quick response.*

Link George agrees on old Engine 11. "She wouldn't run for squat until you were three miles down the road, then it'd be like somebody threw a switch, and she'd take off."

Billy Denny says:

> *My dad got those first trucks after World War II, when things were hard to get. He knew people in the city at International Harvester, and soon as he said it was for a Kent Island fire company they made it a top priority and got those trucks made for us. They were fully equipped with hoses, water tanks and ladders. Ran like a clock. They were tough old trucks.*

Former captain Kenny Bullen adds, "The only thing with automatic transmissions were the ambulances, but that was like driving a hearse."

Curtie Chance, another retired member, knows about the enthusiastic discussion brought to the floor when the purchase of new and expensive equipment is considered:

> *Fourteen was our Class-A pumper. You had to have one to be in the Kent–Queen Anne Association. Eleven was a midmount. When we converted her to a front mount, it took away her Class A. They bought a 1956 Ford, brand new, for about $10,000. Duke kept saying we needed a new truck. The little truck was only 250 gallons.* [Grasonville Volunteer Fire Department charter member] *Vernon Haddaway came to Duke, told him, "Look, we have a demonstrator you can get for a pretty good price." Everybody came to the firehouse, and Vernon brought that Ford 544 truck up there to give us the details. He left. We got to the meeting that night and nobody mentioned the cost. Finally, somebody asked, and Duke said, "Well, they want $30,000 for it." Mr. Josh Bullen cussed out loud and yelled, "That's a thousand dollars a foot!"*

Sometimes the apparatus was bigger than the fire department was ready to accommodate. Prior superintendent of machinery John "Jay" Gibson joined in 1967. He remembers when "we bought that big old Ford truck. I believe it was a '71. You had to load it with gear to back it into the firehouse. Unloaded, it wouldn't clear the doors."

When the discussion turns to advances in communications, former member Bobby Norris says, "I belonged to the fire company before we installed radios in the trucks and started a countywide emergency fire system in 1965."

Ray Baker hurt his back soon after those radios were installed. In bed when the siren went off, his wife asked, "Where are you going?" He told her he was going to the fire. She said, "But you can't do anything." "I can talk on the radio," he answered.

Jay Gibson recalls, "We used to just hear the siren before then. Eventually, we all got monitors."

Calvin Duncan describes how things were done in the early days:

> *If someone had a fire, they'd call Jul. He had a button right there at the store, and he could blow the siren. He had one, and so did Mr. Bill Denny, and they put one in John Holden's house, too. And they had one at the station.*

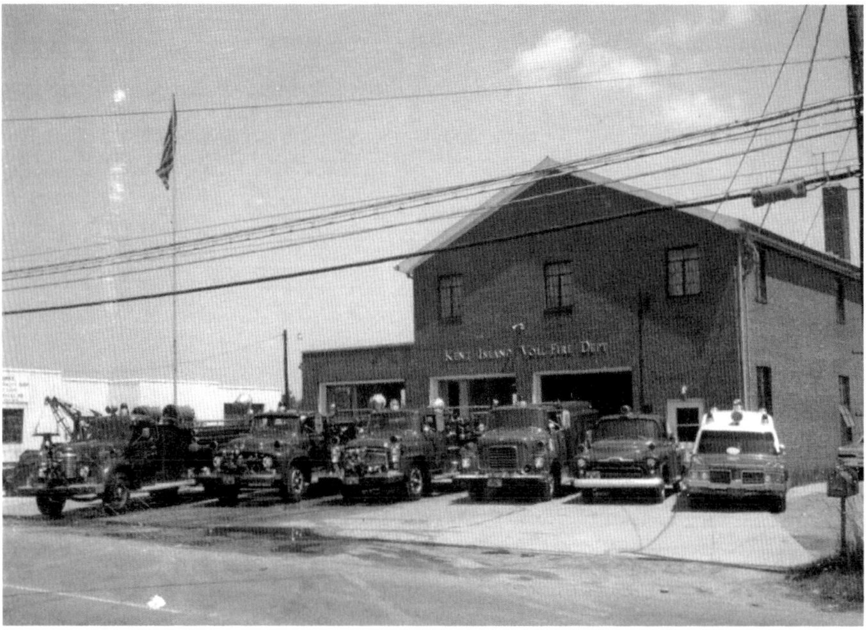

The Oldsmobile ambulance on the far right was totaled in a highway accident. Local watermen were eager to acquire its indomitable 455 engine. *Courtesy Joyce Moore.*

Bobby Norris remembers responding to a siren call:

> *I was picking clams for Joe Buck Baxter. We were sitting down the shore one winter morning waiting to see what the wind was going to do. Robert Horney and I had just joined the fire company. When the siren blew, we took off. We got to the firehouse, and Mr. Denny called. It was a brush fire down Batts Neck. We waited a few minutes, and when no one showed, we hopped in the newest International truck and took off. I was driving; Robert was blowing the siren. We went through Stevensville blaring. I guess we were halfway to Matapeake when Robert said, "Bobby, you do know how to prime the pump to get water don't you?" I said, "No," and panic set in. I said, "Robert, someone has to be behind us," and we actually slowed the truck down a bit. A few seconds later, we could see Juls Grollman's big old Chrysler coming around the bend. We hit the accelerator and siren at the same time in total relief. "Chief" was right behind us. We learned to prime the pump before the next time out.*

As Kent Island grew and the fire department increased its professionalism, new firefighting methods and technologies kept coming along. The *Kent Island News* of August 22, 1963, reported:

> *The Kent Island Volunteer Fire Company just announced* [the] *signing of a contract for the erection of a new radio tower at headquarters which will cost in the neighborhood of $3,000. The local firemen have also voted for safety belts for the ambulance.*

Calvin Duncan offers an example about how new ideas were usually accepted by the leadership of the 1960s:

> *I was working on the Western Shore, at the harbor where the ships unload, and there was a fire in one of the office trailers. Here come the firemen. Man, they jumped off the back of that truck, and they had a one-and-a-half-inch hose on a preconnect. Ran up there and* Pffhht, *they had water. I thought, What the hell is that? We'd roll a three-quarter-inch hose off a reel to fight our fires. I went over and checked it out. Told John Holden about it, told Duke. John said, "No." Duke said to bring it up in a meeting. I did, and they said let's try it on one of our trucks. John was right out there measuring for it. I'll never forget when he was getting ready to put it on. He was drilling a hole. Duke said, "Now, watch it, John. Don't drill too deep in that side there." When he pulled the drill out, water shot out all over the place. He'd driven right through the tank.*
>
> *Before then, we used those old three-quarter-inch lines. You got more pressure from it than a garden hose, but there's only so much water going to come out. They put that one-and-a-half-inch on that first truck, and it soon went on the rest of them.*

Not every new piece of apparatus was greeted with the membership's customary enthusiasm. Calvin discusses an acquisition that needed to grow on people:

> *We got a tanker from Hess Oil Company. We rode up to New Jersey the day we were going to have installation of officers. It was during the '70s gas shortage. We were having trouble getting gas to bring it back. This guy put us wise to a place that would sell us gas. Coming back, we pulled into the station; of course, everybody's there. You know how they are when something new is coming back. It was a green and yellow Hess truck, with*

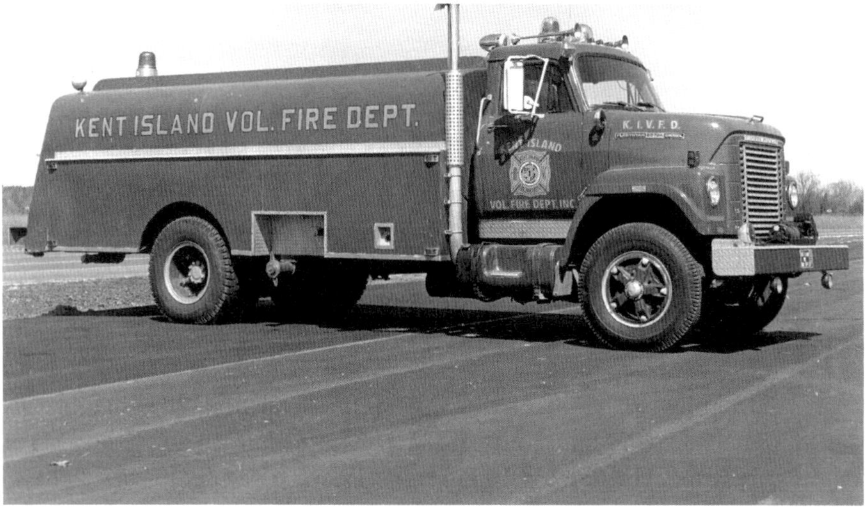

This refitted and repainted three-thousand-gallon 1975 International tanker originated from a donated oil truck. *Courtesy Richard M. Adelman.*

> *that old white rounded nose, and the guys all said, "Put that SOB out back where nobody can see it." And that was our first tanker.*

Jay Gibson recalls, "I painted that truck red. Borrowed a spray gun and painted it out back."

Board member and top responder David "Pops" Gellert Jr. remembers:

> *One night I was driving that tanker up Love Point Road. It was the first time I ever drove it. Lou Johnson asked me, "Well, Dave, you got her to go, can you get her to stop?" I said, "I don't know." I was braking, and it felt like my foot was going right through the floor.*

To illustrate the changing realities of properly equipping a modern firefighting unit, KIVFD bought its first tower truck used in 1999 for a total investment of about $300,000. Though the purchase was a big deal at the time, the transaction was relatively simple. Little more than a decade later, acquiring a new tower truck is a two-year process. First, the chief and members of the apparatus committee set perameters and do basic legwork. Members charged with decision-making then go out and look at various manufacturers. They consult with other departments. When a piece of apparatus such as a new tower truck is purchased, buyers spec it from the ground up. With more hotels, multilevel condominiums and big buildings like Kent Island High School being constructed all the time, a tower truck is a necessity. Chief Paul Schlotterbeck said, "People forget they don't just go up. They go out and over and allow us to get places that would be impossible otherwise." Cost estimates are near $1 million.

That's where fundraising comes in. Former chief Tracy Schulz shares what happened soon after he joined the department in 1978:

> *We all wanted a new fire truck. The newest we had was a '71; you rode on back. Everybody else was getting new ones with cabs. They're safer. Clinton Phillips, the treasurer, was withholding the money because Duke was the only one who could say anything was okay. The old guys asked Calvin how much it would cost. I think it was $135,000. They said, "The only way you're going to get money for a new truck is raise it." I think that was the first time we ever went door to door. Every afternoon, we got on the trucks and went to every community, every driveway and every lane.*

When members past and present talk of equipment in general, everybody agrees with life member Calvin Duncan: "It's very different from the old days. The stuff we had when I was chief, the newest piece was twenty years old. Our gear was old hip boots and long coats."

Kenny Bullen recalls, "If you got to the firehouse and didn't have your gear, you'd use somebody else's. Most gear was kept in a truck. You'd show up and just try to find something. If you couldn't find anything, you'd go anyway."

Life member Jody Schulz says, "We used to get those orange rubber gloves at Grollman's store."

"They'd melt right on your hands," remarks Calvin.

"And we used those old Indian tanks," Curtie Chance says. "Five-gallon metal tanks you'd strap to your back. You had to pump them."

"Basically a big squirt gun," Jody clarifies.

The Indian tanks were used when a truck couldn't get close to the scene. Of one fire in the woods near the Tower Gardens subdivision, Curtie says, "It was dry and really got going. We pulled as much hose as we could and couldn't get any closer. Felt like I carried a hundred of those tanks back there." Today, modern firefighters generally use vinyl bladder bags for this purpose. They're lighter and easier to handle.

With a mission to save the lives of inhabitants of an island, water rescues have often been necessary. There'd be a boat fire or a near-shore sinking. In

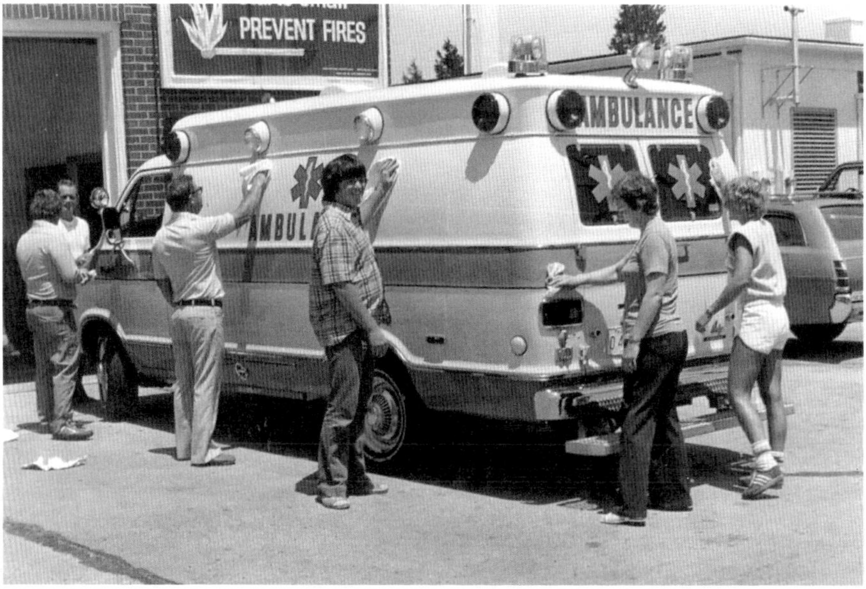

As the 1980s begin, Station 1 members proudly shine up a recent acquisition. *Courtesy KIVFD.*

the winter, hunters would get stranded in duck blinds. They'd go out to shoot birds, and their boat would come loose or sink. They wouldn't come in, and somebody would get worried. "Many of the members were watermen or had other access to boats," says Don Lewis. "Calvin used to have boats, or if he needed one, he'd jump on a friend's workboat, start it and take off."

Don reflects on the changing nature of the Eastern Shore and the KIVFD. "You know," he says, "just a few years ago we still had several watermen in the fire department. Now I don't think we have any."

The company's first official rescue boat was a donated Zodiac inflatable raft. As always in those days, the established leadership was resistant to change. Jody Schulz puts it this way: "The old guys were totally against us getting that boat. Any boat."

Jody's brother, Tracy, says, "A fire made us realize we needed something—a fire on Bodkin Island."

Once a Kent Island peninsula, Bodkin Island is now an eroding speck in the Eastern Bay. Two miles southeast of Turkey Point and uninhabited by humans, what's left of Bodkin provides scant nesting for various seabirds. Tracy Schulz voices the frustration felt on the day of the 1985 Bodkin Island fire:

We couldn't get to it. We didn't know what was going on, if anybody was out there or not. They called a state police helicopter. We flew over, it was just a bulkhead, but a bulkhead that was holding the island together, keeping it from washing away. We had to try and put the fire out. Called the Coast Guard, they came over with a portable pump, we threw everything in their boat, and they'd forgotten the rope to pull start the pump. We had to go all the way back to Southwinds pier to get the rope. We knew we had to do something.

Today, the fire company owns a thirty-three-foot 2006 Northwind Marine Argus class fire and rescue boat with a three-thousand-gallon-per-minute pump that was obtained primarily through Maryland Department of Natural Resources grant money. Boat 1 is rugged and extremely fast. It's docked at a donated boat slip in Castle Marina, which is close to the new firehouse and offers easy access to the Chester River, Kent Narrows and Eastern Bay. The boat is utilized year-round. Going out on the Chesapeake Bay and its tributaries in winter can be especially dangerous.

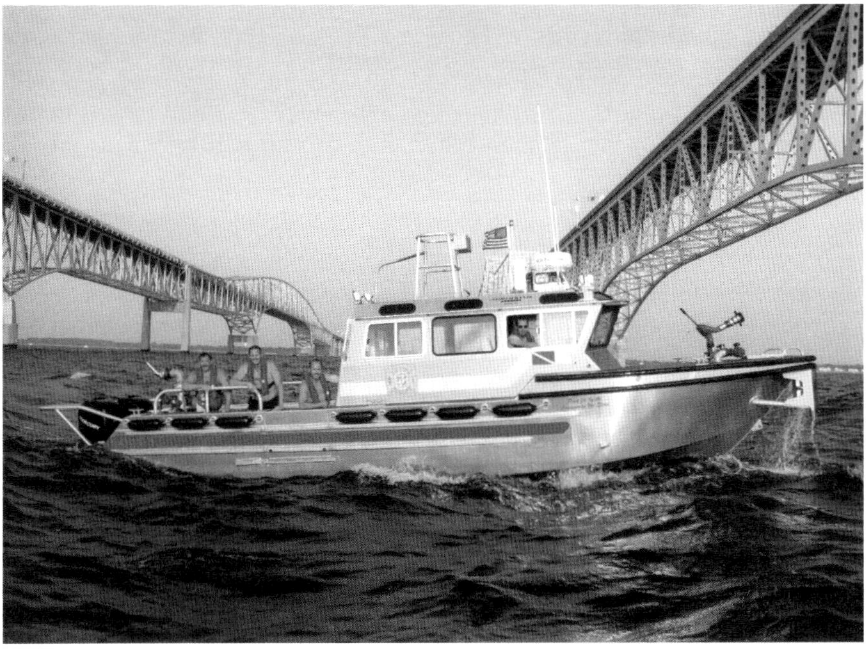

KIVFD's rugged and fast fire and rescue Boat 1 photographed between spans of the William Preston Lane Jr. Memorial Bay Bridge. *Courtesy AsiPhotos.com/Stan Ruddie.*

The department's other boat is a sixteen-foot outboard skiff. Current superintendent of machinery Tom Balderson says, "Six guys can grab that boat and put it anywhere they want."

Today's volunteers hold to professional standards by staying informed about cutting-edge equipment and techniques. They have to think ahead and envision not only the possible emergencies they may face but also solutions to those life-and-death problems.

Former member Bobby Timms addresses the differences between then and now by saying:

> *These guys now are more responsive than reactive. They visualize the future better. Back then, somebody would die; we'd better get this piece of equipment. Gas tanker turns over on the highway, we'd better go buy a foam unit.*

Past chief Paul Fuld notes, "In the old days you had a rack of hose on the back. Maybe a few fire extinguishers. Now it's more professional. It's changed for the better."

Self-described "house mechanic" and member since 1996, Tom Balderson is on the front line of costs considerations. "Today, an air filter for one of these trucks costs forty dollars," he says. "Keeping these big, high-dollar trucks on the road is expensive. I bargain shop, call it smart shopping." Tom's wife, Kelly, is an EMS secretary. "Every ambulance call, any patient care, goes on a medical form. She's part of the department's effort of tracking them through the process for insurance payments. There are a lot of moving parts in this operation."

One equipment issue grew more problematic as the years went on. The KIVFD kept outgrowing its firehouse. The building was added on to a number of times, but that option was eventually exhausted. Doors were changed again and again as trucks got bigger. Eventually, space became so cramped that boat bumpers were used to keep people from banging up the ambulance doors.

In 1988, the *Chesapeake Courier* informed its readers:

> *The volunteers at Kent Island Fire Company are facing a real problem. Their Stevensville firehouse is stretched to the utmost limits of its land. Additional garages have been added on the side and in the rear of their property to house the heavy duty equipment needed to service their mushrooming territory. Now there is no space for one more piece of equipment. Nor is there land to build any more additions.*

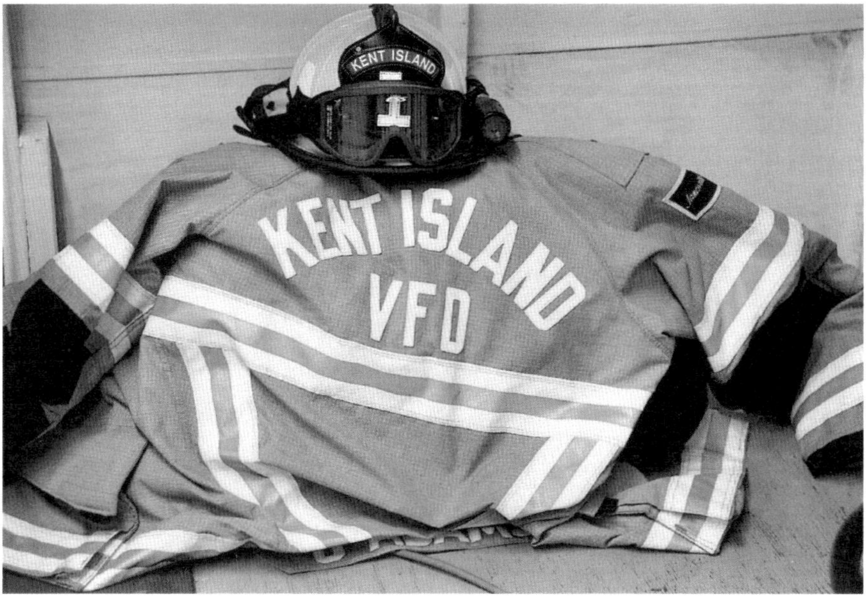

"Bunker" or "turnout" gear refers to a firefighter's boots, pants and jacket, or various combinations of personal protective equipment and clothing. *Courtesy Will Clark.*

In the 1990s, the facility was "busting at the seams," according to Tom Balderson. "Drivers needed to pull trucks out for passengers to get in. We built trucks around the physical restrictions of our house. Apparatus could only be so long and so wide."

In 2002, a 9.6-acre parcel in Chester was purchased by the KIVFD. A building committee was organized, and in 2008, the department moved into its new digs in Chester. The old building that had served the company for sixty years was vacated and sold.

Todd Kerchner, member since the early 1990s, looks at the move objectively. "The old firehouse served its purpose, but we outgrew it. New facilities were absolutely necessary. There are a lot of memories there, but it was time to move on."

The move proved bittersweet for others, however. Soon after the relocation, Dolores Timms says that walking through the vacated firehouse was emotional. "I was there forty years. The new building is wonderful, but the memories are back there."

Kitty Duncan agrees. "It was a sad day when we moved."

Shirley Thomas recalls, "They rang the bell one last time, and we left."

"You could almost feel the memories. See the people," bemoans Dolores. "I cried so hard I sobbed."

It has been said that home is where we love; where our feet may leave, but not our hearts. Sometimes people leave home on fire trucks.

Training

Author, historian and former KIVFD member Nick Hoxter shares the following story about the young department's lack of training:

> *I joined in 1948 or '49. We had that big KB7 International. The siren went off, and I went tearing down to the firehouse. The fire engine was pulling out when I got there. Jimmy White jumped in my car and said, "Turn on your lights and let's follow." We went down past Matapeake to this farm where a real old colonial house was on fire. We jumped out of the car, and I ran to grab a hose. I ran by Josh Bullen when he pulled this huge fire extinguisher off the running board and it went off. Froze my legs, my feet, I thought I was going to freeze to death. They washed me and wrapped me in a blanket. We contained the fire, went back to the firehouse and had a big laugh. But that was when we realized we really needed better training.*

Bobby Norris concedes that the job of the volunteer firefighter in the 1950s wasn't as sophisticated as it is now. "Being first to the firehouse made you the specialist."

Ray Baker remembers occasional training at the University of Maryland's Fire and Rescue Institute in College Park. Visiting expert lecturers and instructors gave members of various area firehouses an opportunity to improve their skills and socialize.

When Ray's son, Keith, joined in the mid-1970s, hands-on training was recognized as essential. He describes one exercise as follows:

> *We'd go in the back bays of the firehouse and pull the trucks out, they'd take all the one-and-a-half-inch hoses and stretch them all out, twist them, intertwine them and put them in loops and knots. Then they'd take your air mask and put tape over the eyes. You'd have to follow that hose all the way from the beginning to the nozzle. When you're in a house on fire, it's dark, it's smoky. You can't see. You have to know by the feel of the hoses, the couplings, the nozzles, which way is out. You only had about twenty*

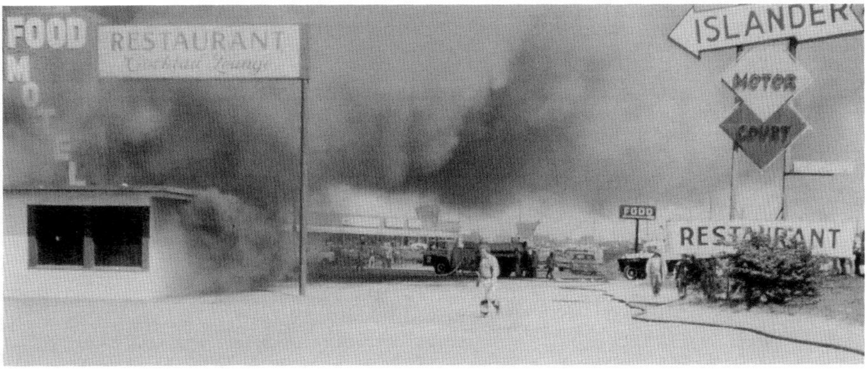

The April 1977 controlled burn of the 1950s-era Islander Inn Restaurant and Motel was one of KIVFD's largest and most memorable. *Courtesy Joyce Moore.*

> *minutes in your air mask. You had to learn how to control your breathing. Some guys got in there and panicked. They'd breathe real fast, and use all their air. You'd hear their warning bells going off.*

Another means of obtaining necessary experience is controlled burns. Owners of uninhabitable residential, commercial and agricultural structures allow the fire department to torch them for training purposes. These exercises give firefighters an opportunity to test their skills without anyone being in any real danger. Since other nearby fire departments typically participate, the controlled burns also help encourage fellowship among volunteers.

For island native Curtie Chance, one regrettable controlled burn sticks out in his mind:

> *It broke my heart when we gutted that old brick mansion over there where Kent Island Elementary School is. That was the old Stevens house, where the Stevens lived when the town of Stevensville started. That spiral staircase was beautiful. A fireplace you could stand in. The place was vacant for years. You couldn't do that today. People wouldn't let you. We used to set marsh fires at the Narrows, too. That kept it from catching on fire later, when they might not be able to control it.*

Despite new breakthroughs in firefighting techniques, Marc Grollman says that even when he joined in the '70s, "the only ambulance training we received was Basic Red Cross First Aid, Advanced First Aid and CPR. It wasn't until later we took classes and became certified as EMTs."

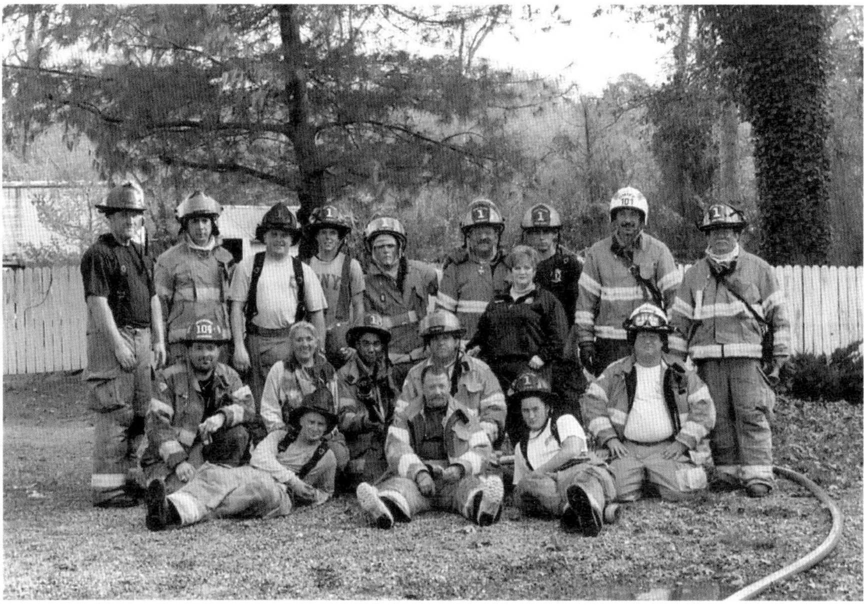

As is traditional, the modern KIVFD consists of seasoned veterans, up-and-coming leadership and new trainees. *Courtesy Will Clark.*

Though the quality of training and levels of professionalism increased over the next couple decades, some members believe that Dougie Thomas's 2001 accident was the real eye-opener. Past president Don Lewis, who at the height of his career answered three to four hundred calls a year, contemplates, "It made you feel all the training you'd had before, that wasn't good enough. It must not have been good enough because Dougie died. So now we have to be even better trained."

KIVFD is also following the national trend of minimizing the desire to be first on the scene. Finance and budget chairman Doug Leatherman acknowledges that haste compromises security. "I think everybody's trying to get away from that way of thinking. In this modern society, it's not safe."

And once the KIVFD responds to an emergency, teamwork is the order of the day. Other companies will be on scene, and as Link George explains, "The bigger the crisis, the more support the department can expect. If you don't work together, somebody's going to get hurt."

Dolores Timms notes how much the job has changed:

Members are far better trained now. Back then you just got on the back of a fire truck, and you went and put out a fire. Now firefighters all have to be certified, they take classes upon classes. Time commitment alone weeds out those who join for the wrong reasons.

A rookie needs two hundred hours of training to get started. Lou Johnson says it's a reality check for recruits:

A lot of them don't realize you're not going to walk in, jump on a truck and go squirt water on a fire. They don't understand the commitment required. First thing, you're assigned to an officer. You're supposed to take the first available fire school. In the meantime you learn every tool and piece of equipment on that truck.

Doug Leatherman describes how a new member can first expect in-house training: "We have a good group of guys who show the junior members around. They teach procedures the new member must learn before they do anything. Then they take a basic firefighting class." Chief Paul Schlotterbeck points out that in-house training prepares candidates for a Firefighter 1

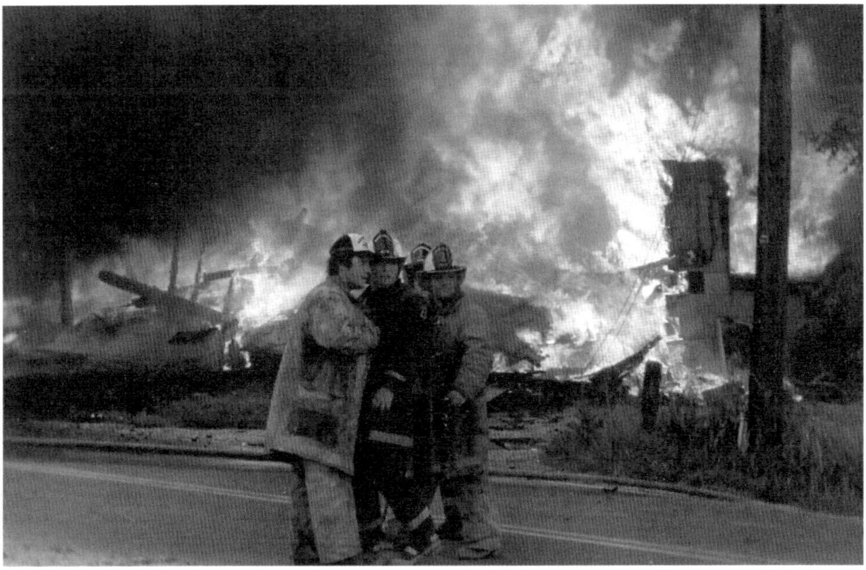

Controlled burns give firefighters an opportunity to acquire experience, hone skills and build camaraderie in a relatively safe setting. *Courtesy Tommy Norman.*

training course approved by the National Fire Prevention Association. He adds, "We, and they, can discover quickly if they're cut out for this."

Captain Bobby Finck comes from a family of volunteers. His mother, Sue Colburn, is a KIVFD EMT, firefighter and instructor. His stepfather, David "Woody" Colburn, has been in the fire service for more than forty years and helped develop the county's Advanced Life Support system in 1987. Bobby says, "Let's be honest. A lot of us join because we want to ride on that shiny fire truck with the lights, blowing the siren and talking on the radio. But that only goes so far. You need more than that to stay in this business."

Pops Gellert fought fires in the navy before he joined the KIVFD in 1987 at the urging of his son:

> *When I got out of the military, I took fire school. It was 60 hours. Now fire school is 131 hours. EMT was 80 hours, now its 141. Firefighter 1, EMT or first responder and a Hazmat course are required. To drive, you have to take courses on pumps and emergency vehicle driving.*

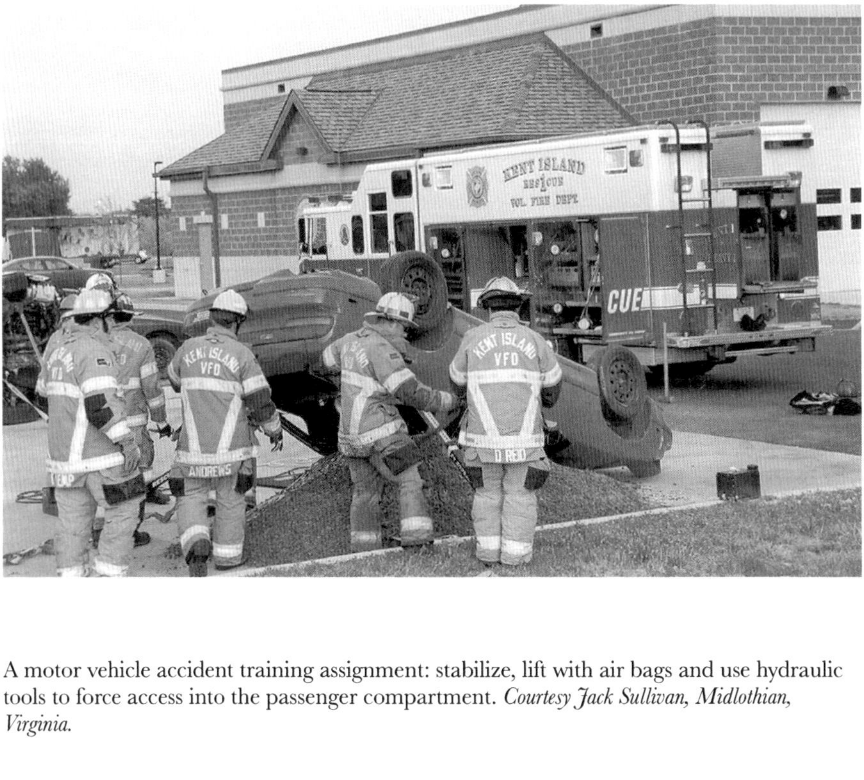

A motor vehicle accident training assignment: stabilize, lift with air bags and use hydraulic tools to force access into the passenger compartment. *Courtesy Jack Sullivan, Midlothian, Virginia.*

Todd Kerchner adds:

> *More and more OPS and CPR training and classes like that are required to stay active. Blood-borne pathogens have to be updated every two years. HIPAA classes, which are patient confidentiality courses, are updated regularly. EMTs and paramedics have to consistently take courses and refresher classes. It's a lot of work just to meet training requirements.*

Chief Paul Schlotterbeck expects training to pay off:

> *I don't think you should be able to walk into a firehouse and know who's getting a paycheck and who's not. The level of professionalism should be the same. These are serious situations we put ourselves in, and we take them seriously.*

Bobby Timms puts all this training in unique perspective. "In most endeavors you train with a hope you get to use those skills. These people train all the time in the hopes they never have to use what they've learned."

Within minutes, a siren is screaming for attention.

Part IV
LIFE IS A FLAME

Life is a flame that is always burning itself out, but catches fire again every time a child is born.
—George Bernard Shaw, Irish playwright

Early Times

Some things never change; others always do. On January 1, 1965, the *Bay Times* carried a story about a Kent Island Fire Department member hunting on an abandoned Love Point Farm. Resting near an old well, the fireman noticed movement at the bottom of the deep, dry-as-a-bone and debris-strewn hole. The newspaper read:

> *Kent Island firemen saved another life this week, but not from a flaming building. A call to the firehouse Tuesday evening received instant action. Chief Moore rounded up a few volunteers and roared to the recue. After clearing away debris at the top of the well, firemen hooked a roof ladder to the edge and fireman Frank Clark made the descent. Picking up the struggling bunny, Clark carried it back up to safety where the rabbit thanked him with a final kick, broke loose and headed for home.*

Forty-five years later, the February 17, 2010 *Bay Times* reported that four KIVFD volunteers—Doug Leatherman, Buddy Thomas, Andrew Calabrese and Bobby Finck—rescued a four-year-old English setter named Buddy that

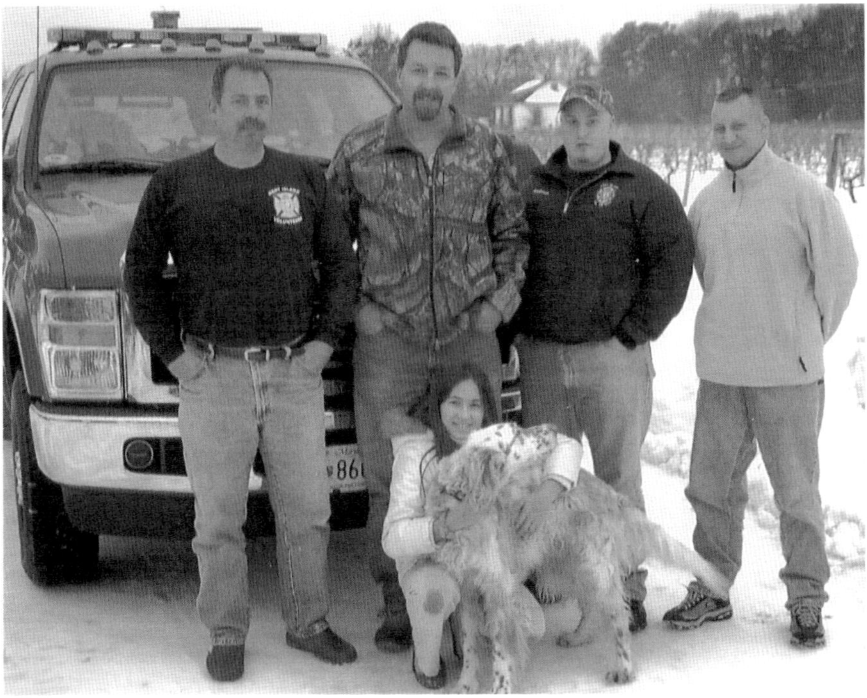

In February 2010, Doug Leatherman, Buddy Thomas, Andrew Calabrese and Bobby Finck rescued Greer Cascia's dog, Buddy, from Thompson's Creek's icy waters. *Courtesy Bay Times.*

had fallen through the ice on Thompson Creek. After finding the dog "stuck in three feet of water, thrashing to stay afloat," the responding team quickly initiated a rescue plan. "Dressed in cold water suits, Finck and Calabrese went on the ice and into the water, freeing the dog," while Thomas and Leatherman waded along the shoreline holding safety ropes tied around the other two men's waists. Once released, Buddy the dog immediately ran home and "slept for about four hours."

Charles Darwin wrote, "The love for all living creatures is the most noble attribute of man." The Kent Island Volunteer Fire Department has embodied this sentiment for over six decades. Another of the firemen's unspoken tenets follows Dwight D. Eisenhower's advice that "a sense of humor is part of leadership, of getting along with people, of getting things done." It's also a way to cope with the occasional unnerving occurrence.

Nick Hoxter recalls:

> *When I was about eighteen, late one evening we were in Denny's Garage when the siren went off. Juls was with us. He said, "Come on, boys, I need your help." We jumped in the car and got to the firehouse, where they said the Silver Roof was on fire. Silver Roof was where Ram's Head is now. We hopped on the back of that big old International with Josh Bullen, John Holden and some others. When we got there, the building wasn't burning, but there were fires all around it. Fire was headed toward Owen Lane's big house over where they eventually built Route 50. I grabbed a shovel, Denny grabbed a shovel, Juls and Josh grabbed Indian tanks and we started fighting fires. It took us about an hour and a half. When we got done, it was too dark to see where we were going. We were wet from walking through the marsh and decided to cut through the cemetery. As we got to the older section, we got separated a little, but I could hear them holler when they'd stumble into a tombstone. All of a sudden, my right leg went through the ground. My foot struck something solid, and I realized I'd stepped thigh high into a grave! It was so deep I couldn't get out. I was trying to pull myself up on a headstone. Man, I was so scared. I hollered, "Help! Help!" Here comes Denny, then Josh and Juls, and they all were laughing. When they finally got me out of there, and we made it back to the main road, Denny asked me if I was really scared! I said, "Are you kidding? I'm still shaking." Juls and Josh never let me forget it.*

Billy Denny just says, "I couldn't believe what I'd seen."

"I lost that shovel that night, and for all I care it could still be there," Nick quips.

Life member Jay Gibson remembers another incident:

> *One time Calvin* [Duncan] *and I were called to a house fire. For some reason we didn't take the little truck, we took 14, that '56 Ford. Calvin was driving. We pulled into Sunny Isle of Kent, and we couldn't see any smoke or fire or anything. This gentleman was standing off by the side of the road, and we asked him where the house was. He walked to the side of the truck, came up with a pistol and pointed it right in my face. He said, "Get out of here. Back this truck up, and get out of here." Calvin hits first gear instead of reverse and I thought, well, that's the end of me. We managed to get out of there though, and called the state police. I don't know where the trooper came from, but he got there in a big hurry and disarmed*

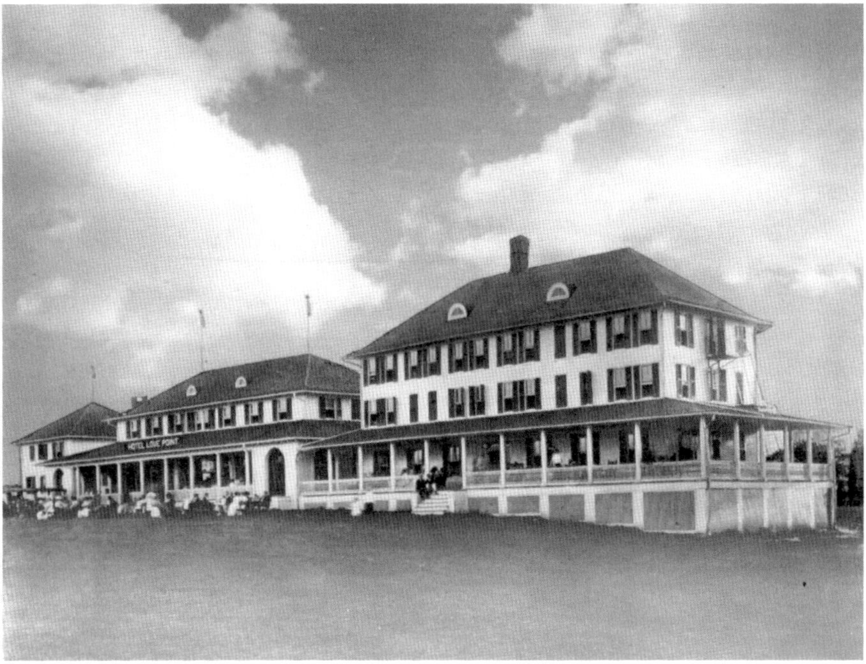

Hotel Love Point was a massive symbol of early twentieth-century grandeur and Eastern Shore hospitality. *Courtesy Michael S. Kader, Wye Knot Prints and Collectibles.*

> *the guy. Somebody said the guy's wife wouldn't fix him lunch, so he doused the place in gas.*

Calvin Duncan chuckles. "That man was having marriage problems."

It wasn't always gallows humor and sidearms. Sometimes KIVFD members watched history burn.

In the early twentieth century, Kent Island's northern tip was the site of the famed Hotel Love Point. This elegant and sprawling landmark resort was serviced by the great Chesapeake Bay steamboats. Vacationers and day-trippers arrived from Baltimore by the thousands to enjoy an Eastern Shore paradise. Guests fished and crabbed, swam from the beach and visited the amusement boardwalk. Cottonwood trees shaded the vast picnic grounds. Post-Victorian romantics canoodled in canoes at the lake. Beyond an expansive lawn, looking down from its hilltop perch, sat the massive hotel with its broad veranda and wraparound screened porches. A majestic dining room served abundant traditional Eastern Shore fare to out-of-towners and locals alike. Saturday night dances were also popular with both crowds.

Years of behind-the-scenes financial woes were exacerbated by the beginning of the Great Depression, and the hotel closed. Several attempts were made to reopen and revive the excitement of the heady years, but times change. The age of the Chesapeake Bay resort was ending if not already over. The once opulent property commonly referred to as the Love Point Hotel sat abandoned for decades, until November 12, 1965, when a suspicious fire broke out.

According to the *Bay Times*:

> *An Eastern Shore landmark went out of existence in a blaze of fury Friday afternoon when the old Love Point Hotel became the victim of what police believe was arson and burned to the ground. The blaze started a little after one o'clock in the afternoon in the northern end of the structure and rapidly climbed to the roof where it quickly spread the entire length of the building in minutes. The tinder dry old wooden hotel soon became a mass of flames with smoke billowing to the sky.*
>
> *The Kent Island Volunteer Fire Company arrived on the scene to find a flaming inferno and after a few seconds attempting to save part of the building concentrated their efforts in keeping the flames from spreading to adjacent areas. The heat from the fire was so intense that firemen donned reflective outer garments and spectators shielded their faces as they watched. After the fire was brought under control, the Island firemen found that the fire engine dome lights had melted under the intense high temperatures.*
>
> *About 20 minutes after the hotel caught fire the building started to collapse, shooting sparks skyward as flame ate away its support members. In 30 minutes, nothing was left but a pile of smoldering rubble and two chimneys outlined starkly against the skyline.*
>
> *Tendrils of smoke still laced patterns into the air from the ashen bed of the old hotel as late as Wednesday for the hundreds of countians who drove up for a last look at the Eastern Shore's Palace of Pleasure and reminisce about the good old days.*

Billy Denny rues the day. "When I heard that hotel was on fire, I went up there and it was engulfed," he says. "All the way, not like fire usually is in spots. All three big sections were in flames. I shook my head and said that's not good. It was the middle of the day and there was nothing to cause it. That was a real shame."

Babe Grollman agrees. "That was a huge, beautiful building, and it did burn."

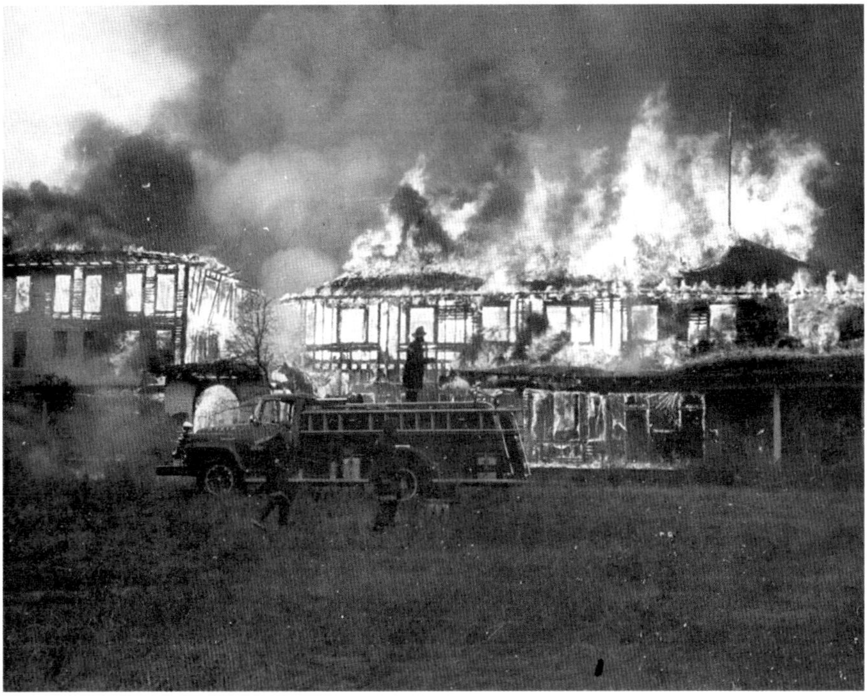

According to Calvin Duncan, "Stumpy Grimes pulled the brush-truck too close to the Love Point Hotel fire and melted the beacons off." *Courtesy Kent Island Heritage Society.*

Kent Island is separated from the Eastern Shore mainland by Kent Narrows, a small strait with a strong current. The Kent Island side of the Narrows is home to the Kent Island Yacht Club, county slips intended for a dwindling workboat fleet, the Chesapeake Exploration Center, a few residences and a scattering of commercial properties, including a marina. Future development is looming.

The Grasonville side of the Narrows is now known for its landmark restaurants, popular hotels, upscale waterfront condos and marina space to fit every boating niche, but for half the twentieth century, Kent Narrows East was a hub of Maryland commercial fisheries. At one time, there were as many as fourteen active oyster shucking and packinghouses here. Most employees were transient and lodged in nearby tinderbox shanties with no utilities except electric. The Narrows was where most of these businesses were located, but there were a few in the Dominion and Little Creek areas of Kent Island as well. Eventually, the local seafood trade began to collapse, and as the community's economic eye turned toward tourism, the shanties were more of an eyesore than ever. The last was razed in the 1990s.

Bobby Norris recalls:

> *Seemed every winter there was a shanty fire. If you'd ever been in one of those shanties, the walls were lined with cardboard and newspaper, and they all had wood or oil stoves. If the wind was right, the marsh would catch on fire.*

One of Kenny Bullen's first on-scene experiences was a Kent Narrows shanty fire. "They had big LP tanks," he says.

> *I was standing there spraying the fire when the safety valve popped on one of those things, and it was like a giant blowtorch. Shot right straight across to a second shanty and lit the whole side of it up. I took off. Duke Moore kept hollering, "Come back with that hose! Come back with that hose!"*

Much of the Queen Anne's county seat of Centreville was added to National Register of Historic Places in 2004. On April 25, 1968, however, the *Queen Anne's Record-Observer* reported:

> *Three buildings in the heart of Centreville's business district lay in ruins today after flames threatened much of the town about 9 o'clock Tuesday*

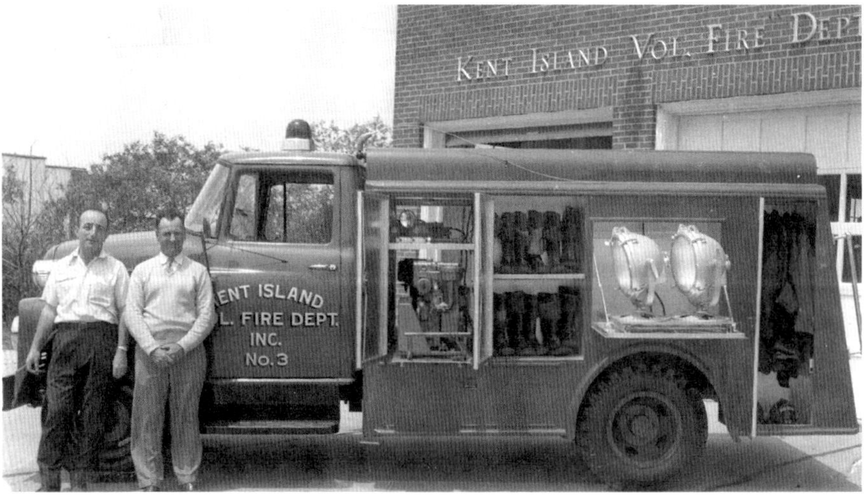

Charter members Julius Grollman and Orem Thompson pose with Engine #3. *Courtesy KIVFD.*

> *night. The fire, the third disastrous blaze to strike the town's business area since 1902, completely destroyed the old, long vacant, Opera House building, Thompson's Pharmacy, Mayer's Shoe Shop, and the Beauty Spot Salon. Smoke and water damage occurred in Center Coffee Shoppe, The Center Beverage Store, and Draper's Market all on Water Street. These stores were burned out on July 11, 1963, in a similar holocaust.*
>
> *Firemen were getting the first hose lines into operation when there was a muffled explosion and the entire roof of the old building erupted into flames.*

Calvin Duncan was on scene. "The Centreville Opera House fire was a big one. There were high winds. Cinders coming off that theatre were setting other roofs on fire. They could've lost the town that night."

The *Record-Observer* story continued:

> *From a spectator's point of view it was almost miraculous that firemen were able to keep the flames from spreading south on Commerce Street into the Blue Lantern Inn and other old frame houses fronting on the Main Street. The firefighters also did a tremendous job of stopping the flames that spread out of the long Opera House building into the rear of the packaged goods store and Draper's Market. At the height of the blaze, sparks were flying over most of Centreville and at least four separate fires started on roofs within a block or two of the main blaze.*

Curtie Chance was there, too.

> *When we drove over the Narrows Bridge you could see the fire in the distance dancing in the air. She was a hot one. We were on the roof of Draper's Market, across from the old Barton Brother's Funeral Home. When your relief got there, boy, you were ready to go.*

Sometimes in the chaos of an emergency situation, first responders are so focused on the job at hand that details of a victim's identity may elude them. Past president Ray Baker remembers the day in 1961 when Kenny Bullen's uncle, William Bullen, was killed:

> *We pulled up, and there was a person lying on the highway. The state trooper told us we were too late to help him. We went over by the truck where Kenny's aunt was. Good lord, it was bad. I never thought she'd survive.*

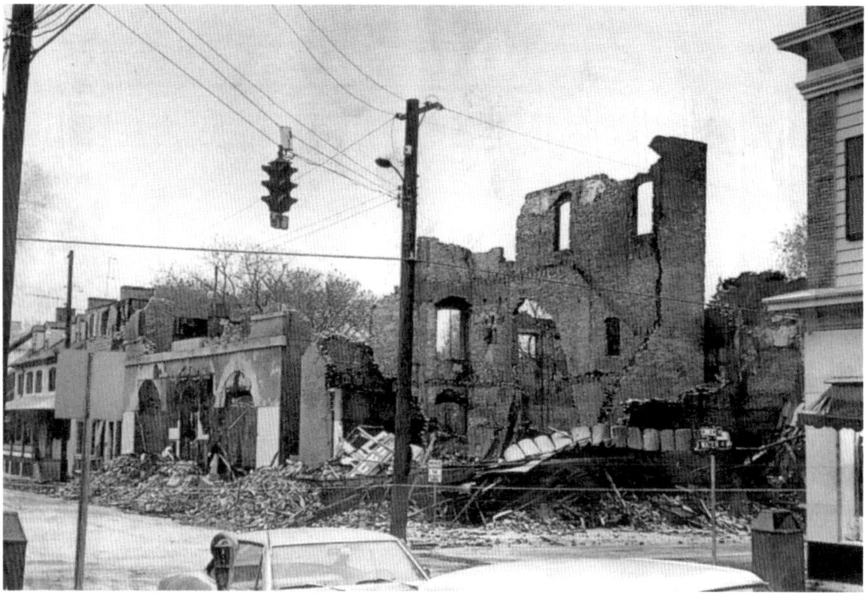

Fire destroyed a number of buildings in the heart of Centreville on April 23, 1968, including the historic opera house (circa 1907). *Courtesy Woody Woodford.*

Kenny Bullen recalls, "I was on the ambulance and couldn't even tell who it was. They were in Earl Seward's truck, so I didn't recognize it either."

Ray Baker says, "I started to realize who we had in the ambulance on our way to Easton. The main thing was to get her to the hospital. I didn't tell Kenny until we got back."

"There was a crossover this side of the Narrows," Kenny continues, "and they were using it when somebody struck them. Ray, Duke and I responded. I had no idea who it was until afterwards."

Mrs. Pauline Bullen survived her injuries to live until 1996, when she died at age seventy.

Island native Jody Schulz explains, "That happens. You're hoping you don't know them to start with." He considered why this occurs. "Maybe you put a block up there."

Ray Baker adds, "You're too busy taking care of the problem at hand. People ask, 'What kind of car was it?' and I'd say, 'Damned if I know.'"

And occasionally there's no doubt that you know the victim of an accident. Ray's son, Keith, shares the following:

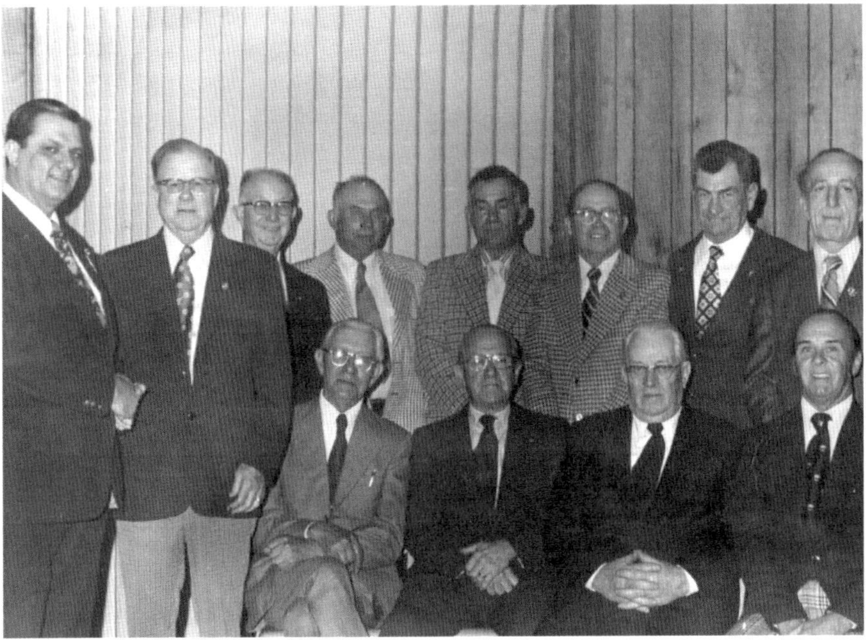

Unknown state official, Josh Bullen, Lee Bell, Ellwood Smith, Babe Collier, Billy Lane, Francis Price, Julius Grollman, Pat Thomas, Bill Denny, Claude Lowery and Billy Thomas. *Courtesy KIVFD.*

> *John Cornwell was our age, a friend of everybody in town. One day he was hunting, and his gun discharged. Shot himself right in the chest. He had a fifty-cent-piece hole there. Duke and I were on the call. I was in the back of the ambulance with John, I talked to him, treating his wound the best I could. There weren't any IVs in the back of the ambulance then. We took him to the hospital in Annapolis. All the way over there, he kept saying, "Keith, Keith, man. I don't want to die. I don't want to die." I told him, "We're going to get you there, John. Everything's going to be good." He died at the hospital. It was probably my second ambulance call. It was a while before I realized I might have been the last person he saw that he knew.*

As in any community, arson is an occasional ugly presence. Curtie Chance recalls a rash of island pyromania:

> *One woman was going around setting stuff on fire. A man named Dillihunt was the fire marshal. Somebody took her to Cambridge* [the location of a facility for people with mental illnesses then known as the Eastern Shore

> State Hospital], *and they let her out. We saw Dillihunt and told him we'd seen her car. He said, "Aw, no, I told them to call me if they let her out." Two or three guys kept telling him she was running around here; finally he called, and they'd let her out two weeks ago. Man, she was back out lighting fires.*
>
> *There was Miss Edie Palmer's farm. You go down there towards Turkey Point, what they call Southwinds now, that little farm there on the left, about forty-five acres. Ran along the water a good ways, there's four or five houses there now. Gibsons lived there then. It was probably a tenant house. I was picking corn, headed south, and saw all this smoke. I stopped the combine and got out, drove the old farm truck there. That's the day the Keys' house burnt up. They figured she'd set that, too.*

The October 30, 1969 edition of the *Bay Times* detailed how the Benton's Pleasure home of Francis Scott Key, descendant of the composer of "The Star-Spangled Banner," was destroyed by fire and that "Grasonville [volunteer fire department, established 1958], Queenstown and Centreville sent units to assist when a shortage of manpower and water hampered efforts to stop the blaze [at] the Benton's Pleasure home."

"And there was that old Russian guy's trailer up Love Point," Curtie recalls. The *Bay Times* of November 11, 1969, reported:

> *An elderly Russian born recluse who has lived at Love Point on Kent Island for the last 20 years lost everything but the clothes on his back when fire broke out in his home Saturday. The fire erupted about 10:30 a.m. in the home of Carl Kozeduk, 78 years, which was a tin covered shantie in the back of the old Philmore Hotel. Firemen found the hut collapsing in flames when they arrived and could do little more than prevent the spread to other buildings while attempting to keep the old man from entering the structure to save his animal friends.* [Several cats and rabbits Kozeduk kept for company.] *Kozeduk sat outside the embers of his home for hours weeping tears of sorrow long after firemen had secured and left the scene. In gratitude to them he tried to give away the lone possession he had besides his clothes, a sledgehammer that had somehow escaped the flames. As word spread of the old man's plight, the use of an old railroad car was offered. Soon clothes were brought and a bed was found for the destitute man. Monday the electricity was turned on and Kozeduk had water and light, and an indoor toilet, luxuries he had long gone without.*

"Talk about arson," says Jody Schulz, a five-year veteran NFL linebacker for the Philadelphia Eagles. "Halloween used to be wild, especially down

Dominion." Dominion, formerly Devil's Dominion, and nearby Little Creek were hardworking watermen's enclaves where rowdiness was not unheard of. Tracy Schulz recalls:

> *They'd set fires back one of those roads; you'd respond, start heading back there* [and] *they'd throw eggs at you. One Halloween we had twenty-some fires in three days. We'd take the brush truck because the engine was brand new, and Calvin said, "Don't take that truck down there." They set the church on fire, the junkyard. It was crazy.*

Buddy Thomas grew up on the island. "Cops would go back there to get the people setting fires, and others would be on the roofs pelting them with eggs. The police would chase the people throwing stuff and the guys setting the fires would get away."

Jody remembers an arson fire in which the culprit, a new member of the fire department, unwittingly gave himself up.

> *I was suspicious when he was* [the] *first one at the firehouse. He was never first. Then we were on back of the truck on our way to the fire, and he yelled, "She's really burning!" I wondered how do you know she's burning? After we got it under control, I thought I better tell somebody.*

In an unrelated incident, the *Bay Times* reported on January 15, 1970:

> *A stubborn fire of undetermined origin destroyed the historic manor house of Blakeford, near Queenstown, Tuesday. Firemen from 12 companies failed to stop the blaze that is believed to have originated in the basement of the home built in 1834 and raced up the walls and spread across the roof. The first notice of the blaze came about 7:40 a.m. By 3 p.m. it was nearly all over. Firemen hooked winches to the weakened walls and pulled them to the ground for safety's sake. Well into Tuesday night fire trucks and men were at the scene mopping up lingering blazes springing to life in the jumbled pile that was all that remains of the fine old home.*

Calvin Duncan recalls:

> *Blakeford was a big fire. It took fourteen hours to fight it. Cape St. Claire was called from across the bridge to help, then Earleigh Heights, but not*

> *Kent Island. Chief John Holden was so mad; he sat there in that station boiling. Finally, at eight o'clock at night they called us because they needed lights, and we had a light truck, Number 13, a big International. Duke and I get there;* [Queenstown's first assistant chief] *Vince Anthony shook our hands and said thanks for coming. He got along good with Kent Island. We always kidded each other. I said, "Vince, if you're ashamed of us, put us around back where nobody can see us." He laughed. I told him, "As long as you all took to call us, Kent Island could have blocked and bricked the doors and windows and filled the building up with water by now."*

As the younger generation became more active in the 1970s, safety was sometimes compromised by efforts to respond quickly. Marc Grollman says:

> *You needed to be careful responding, especially at night. Keith Baker, Bobby Timms, me, Calvin, Duke, maybe a dozen people coming from Cloverfields and Stevensville all at one time, going fast, and headed toward one spot.*

Without providing details, he admitts, "There were incidences."

Once they'd arrived on the scene, things didn't always go smoothly, either. Dolores Timms remembers a residential fire her son, Bobby, responded to in Cloverfields. "The owners were away," she says,

> *but the neighbors told the boys there was someone inside. Calvin was chief so he sent Bobby and Bruce Coursey in to see if there were any people in the house. Its pitch black in there, dark and smoky. They're on their hands and knees, and Bruce yells, "I got him! I got him!" They grab hold, pull and pull...and drag out a beanbag chair.*

No one was inside the house.

Changing Times

Summer evenings on the Eastern Shore are muggy and slow. In the mid-1970s, they were usually also quiet. But early one Saturday night, halfway through a decade that brought enormous change to Kent Island, the peaceful twilight was shattered by violence unimagined in this small but growing community.

On August 9, 1975, thirty-nine-year-old Maryland State Police sergeant Wallace J. Mowbray took an on-duty cruise around the island. He stopped to see his wife and youngest son at home in Chester for a few minutes before continuing his rounds. He didn't get far. He had a reputation for being a natural at his job, so when he saw a blue van backed into a shadowy section of the Baker's Liquors parking lot, it caught his attention. He drove up next to the suspicious vehicle, which had, in fact, been stolen.

Sergeant Mowbray exited his police car and approached the van. The driver gave the state trooper a license but claimed not to have a registration. As Mowbray took a look at the Ford's out-of-state dealer tags, the four young Western Shoremen inside made a decision that would irrevocably change countless lives.

Wally Mowbray returned to his car. The driver of the van, Albert White Jr., leaned back in his seat. Richard Patterson reached over from the passenger side and aimed a sawed-off shotgun out the window.

Charles Quimby, working at the Kent Island Mobil station across the street, told the *Queen Anne's Record-Observer* that "after Trooper Mowbray got back in his car, as he got in there, the shot went off."

Quimby and other witnesses, including future KIVFD members dining at the popular Tastee Freez drive-in that occupied the western corner of Cox Neck Road and Routes 50/301, watched as the van sped away, ran the traffic light at the intersection and disappeared out of sight.

The *Record-Observer* went on to report that Mowbray was "treated by a Stevensville ambulance crew and flown to Baltimore's Shock Trauma Unit by state police helicopter."

Keith Baker responded to the call. Thirty-five years later, he discusses the events of that night:

> *Duke and I were sitting in the firehouse radio room office like we had so many times, listening to chatter on the radio. It was still light out. Duke was at his desk with his feet up; I was sitting on a bench. He leaned back listening and heard something. He said, "We're getting ready to get a call here." Next thing you know, the fire board's alerting Station 1 an officer's been shot. It was one of my early calls; I didn't have a lot of experience. Duke was reluctant about taking just me with him, but he said, "Don't worry; there will be somebody there that can go with us."*
>
> *When we got there, it was a crowd but nobody that could help. We got him in the ambulance. I was in back with him all by myself. We left, and I started doing CPR. I couldn't feel a pulse. That's when I looked down, saw his nametag and realized who it was. Duke called the fire board for a helicopter, but in those days I think the state only had two helicopters and*

> *neither was available. We headed to Easton. Duke called ahead to Station 2 at Grasonville. They had an attendant standing by waiting for us to pick him up. Almost as soon as he got in the ambulance, the fire board called back and reported the helicopter was on route to Bay Bridge Airport. Duke said, "That helicopter better be there," and he headed back to the island. We got to the airport in less than five minutes. Soon as we pulled in, we could see the helicopter coming down. We helped load him in and watched them take off. My white T-shirt was soaked red.*

"It was one of the worst things that ever happened here," says Kenny Bullen. "Nobody could believe it."

Ladies auxiliary member Cindy Taylor is succinct: "It took our innocence away."

Keith Baker continues:

> *Cops were everywhere. Helicopters were flying around. These guys who murdered a state trooper were running loose. I was home alone and freaking out. I went over and talked to Juls Grollman a while. I talked to* [Stevensville resident] *Buddy Wilkins. I headed towards Castle Marina to see if my father was at his boat. I was driving too fast in my '63 Chevy II and turned it over in Bus Tanner's soybean field. The police came, and while I was sitting in the car a bunch of onlookers were saying I shot Mowbray. It seemed strange because I was one of the few people with him after the shooting.*

The fugitives abandoned the van in Stevensville and split into pairs. Albert White and the triggerman's half brother, Raymond Patterson, stole a boat and crossed the Chesapeake Bay. Richard Patterson and Charles Watson, who according to court records encouraged the killing of Sergeant Mowbray from the back of the van, accosted a man and woman in the parking lot of a bay-front restaurant.

Brandishing the shotgun, Watson and Patterson forced the couple to drive them around the Eastern Shore for four hours. When the kidnappers finally demanded to be transported across the Bay Bridge, Patterson raped the female while Watson kept the weapon leveled at the male driver.

Police officers stationed at the bridge's westbound terminus noticed the small foreign car with four occupants as it neared the tollbooths located there at the time. When they approached, the driver suddenly yelled, "They've got a gun," knocking its barrel away from his head. After a brief struggle,

Watson and Richard Patterson were taken into custody. Raymond Patterson was apprehended later that night walking through an Anne Arundel County residential area. White was arrested on Route 95 near Perryville, Maryland, soon after.

And not quite four hours after suffering a massive head wound at point-blank range, Wallace Mowbray died shortly after midnight on August 10 while undergoing surgery.

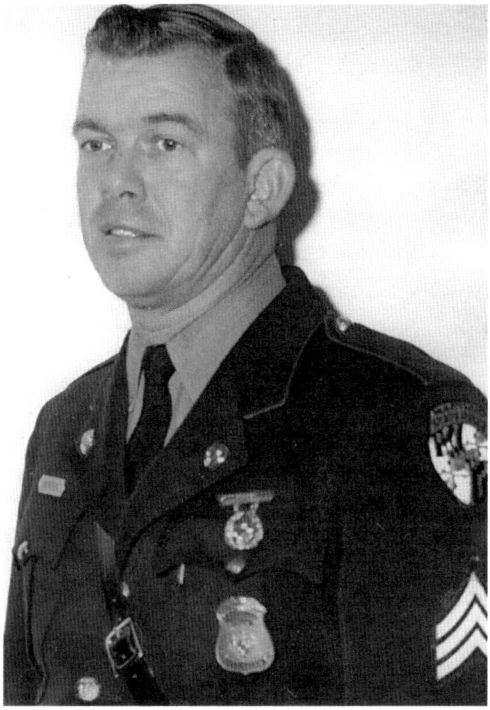

Wallace Mowbray was a well-respected family man, community anchor and Maryland State Police officer. His 1975 murder shocked the region. *Courtesy Mowbray family.*

Over one thousand people attended Sergeant Mowbray's funeral the following Tuesday at Chester's Kent Island United Methodist Church, mere yards from the scene of the crime. The church was overflowing. Hundreds of mourners, including a multitude of solemn police officers representing departments from across the Mid-Atlantic, stood on the lawn and paid their respects, despite oppressive heat and humidity. Governor Marvin Mandel and other state officials were in attendance. Reverend Roy B. Phillips, church pastor and Kent Island Volunteer Fire Department chaplain, co-officiated with Father Charles K. Riepe, director of the state police chaplain's division. Lieutenant Marlin Clevenger, one of Sergeant Mowbray's best friends, told the congregated, "This man had more natural police ability than any man I've ever seen. Yet he had a lot of compassion for people. He believed in justice, the right kind of justice. Justice tempered with a little mercy."

KIVFD served coffee and desserts to visiting law officers before the funeral. Dolores Timms recalls looking out the firehouse's upstairs windows at a "sea of police cars."

An editorial in the *Record-Observer* memorialized the fallen trooper:

> *Every citizen of Queen Anne's County has his or her own personal recollections of Wallace Mowbray—not only of our contacts with his work, but more importantly with him as a citizen, as a neighbor, as a person whose private life was a mirror reflection of the principles by which he lived in carrying out the law which he had sworn to enforce. The very existence of these personal recollections is a more eloquent and lasting tribute than anyone can give, for the character of such a man is so complete that it neither requires nor permits definition or embellishment. No less can be said of his professional life.*
>
> *It involves no disrespect to his fellow officers to say that less than a handful were even nearly his equal. With even more certainty it can be said that none were his superior.*
>
> *It's too late now to say "Thank you, Wally Mowbray." But God grant that we won't forget your example.*

Three of the four assailants are now deceased. Richard Patterson remains incarcerated.

Trooper Mowbray's family—wife Iris and children John Russell "Rusty," Robin and Ryan—grapple with the repercussions of their loved one's murder to this day. Robin, currently living in Cambridge, Maryland, speaks for her mom and brothers when she talks about the support provided by the residents of Kent Island in the tragedy's aftermath. "We had wonderful neighbors," she says. "Everyone wanted to help us. People were so kind and respectful. That they were there for us during that time of our lives means so much to all of us. It touches your heart."

A new era of emergency response had begun. Violent crime became a concern where little had existed before. Current KIVFD president Jody Schulz recalls:

> *Not long after I joined the department, Dougie* [Thomas] *and I responded to a reported stabbing in Dominion. When we got to the house, we recognized a Monte Carlo in the driveway with tags that read "Dr. Love." Dr. Love was this one-legged guy we all knew, a real character. A lady at the scene said a woman on the third floor with a knife was after Love.*
>
> *Dougie and I go up these little narrow steps, and there's Dr. Love lying on a mattress. This woman's standing over him holding a butcher's knife. He had a couple cuts on his arm. It must have been 110 degrees up there.*

Dougie yelled, "Put down the knife you crazy so-and-so!" Cap, she turned and looked at Dougie and said, "I'll cut you like I did him." We heard the cops pull up outside. I said, "Don't worry, Love, the law's here."

We passed the cops at the front door. On the way out, Dougie just pointed up and said, "Dr. Love and crazy woman with a knife, third floor."

Tracy Schulz remembers, "One time this man down Little Creek was messing with another woman. His wife broke in through a window. The girl got out, but the wife stabbed him."

"He wrapped himself in a blanket and staggered out," continues Jody, "then dropped dead in this big pool of blood. After we loaded him into the ambulance, Norman Bodani sat in back lecturing this corpse. 'Look at you, look at you. All over a woman. Was she worth it?'"

A traffic accident grabbed everybody's attention on March 13, 1980.

That Thursday morning was miserable with the sharp winds and freezing rain often associated with late winter on the Chesapeake Bay. North Carolinian Ronnie Creason, driving an empty livestock tractor-trailer on the Kent Island end of the Bay Bridge's westbound span, struck an icy patch and lost control of his rig. The truck went into a skid, slammed into the guardrail on the

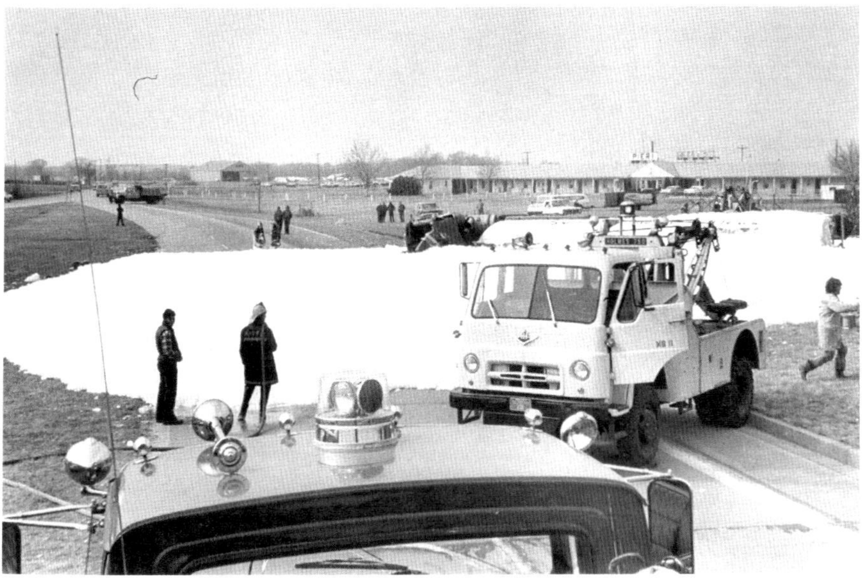

In March 1977, the brakes failed on commercial truck, spilling six thousand gallons of highly flammable liquid. A curious passerby asked, "What's happening?" while smoking a cigarette. *Courtesy KIVFD.*

driver's side and then swerved and crashed into the guardrail on the right. His cab stopped, but his trailer didn't. Creason's problems were just beginning.

The trailer pushed the cab through the guardrail, and for the next hour or so the cab of the truck swayed, suspended fifty-feet over the cold and choppy bay with the driver stuck inside.

Rubber brake hoses were all that kept man and cab aloft until witnesses were able to secure the truck to the bridge using a few links of chain and a car jack handle. When the authorities arrived on the scene, these good Samaritans, mostly unidentified, went on their way.

Authorities included members of the Kent Island Volunteer Fire Department. "We asked for volunteers to go get him," says Link George. "There weren't many hands raised."

Attempts to use a fire ladder and then a Stokes basket to rescue Creason proved ineffective. Then Lieutenant Bruce Coursey remembered a nearby construction site where there was a heavy-duty crane and basket. The owner gave Bruce permission, and the equipment was moved to the scene.

The *Bay Times* summed up the event's conclusion as follows:

> *Duke Moore, Assistant Chief, an Island fireman and bridge workman, along with Coursey climbed into the basket which was lowered to the door of the swaying cab. The door opened and with Creason safe, the truck was lifted and lowered to the surface of the bridge. Creason was taken to Anne Arundel General Hospital followed by an entourage of newsmen, where he was treated and released.*

Later that year, a fire was the talk of the town. The Kent Narrows restaurant Fisherman's Inn was originally owned by Captain Alex Thomas, father of Harold Thomas and Betty Thomas Schulz. A business that started out with six tables in 1930 has, under the leadership of Betty and her husband, Sonny, along with their three sons, Andy, Jody and Tracy, expanded today into a busy complex of two popular family restaurants and a seafood market.

"Fisherman's Inn burned down December 23, 1980," says Jody Schulz. "Tracy and I were on one of the fire trucks when we saw this car flying up the highway behind us. We looked over as the car went past, and it was our mother."

Marc Grollman grew up with Andy, Jody and Tracy Schulz:

> *We used to get calls to Fisherman's a lot. The sign was on fire, or something in the kitchen. That night they said Fisherman's Inn was on fire, we hardly*

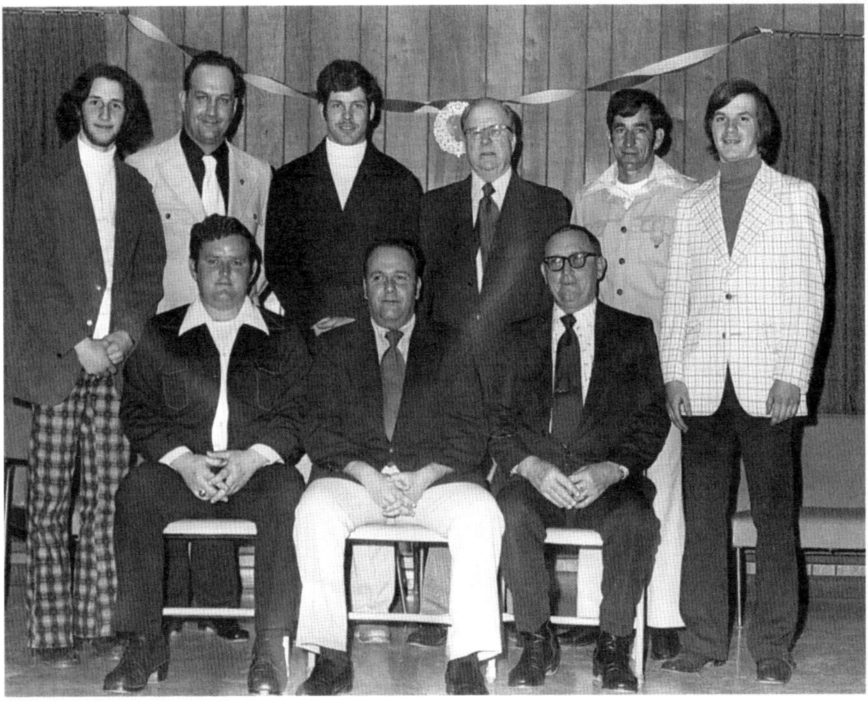

The Mod Squad. *Top*: Marc Grollman, Link George, Byron Phillips, Josh Bullen, Calvin Duncan and Keith Baker. *Seated*: Billy Moore, Jay Gibson and Duke Moore. *Courtesy KIVFD.*

> *thought anything of it. When we went over the Narrows Bridge, we saw the flames and smelled the smoke; we went, whoa, Fisherman's really is on fire.*

Buddy Thomas is first cousin to the Schulz brothers and one of the most respected members of the current KIVFD leadership. His wife, Colleen, is an EMT and works in the department's billing office. Buddy says, "We were on the roof. Calvin started hollering to get off. He was last off and right behind me. I was climbing down and he was standing on my helmet."

Bobby Timms was close friends with everybody there. "We were stunned, trying to get our equipment off the roof. Calvin threw the saw off and was yelling—" Now a lay minister in his church, Bobby pauses. "Well, I won't tell you what he was yelling."

Oscar "Sonny" Schulz, onetime waterman and longtime civic leader who later served as Queen Anne's county commissioner, watched the business he and his family worked so hard to build burn. "It was so cold the ladder truck froze. When they made a hole in the roof, the windows exploded."

"A bunch of us were inside fighting the fire," says Marc Grollman. "Some were on the roof. Calvin Duncan came running in yelling for everybody to get out quick. Just leave the equipment. Seemed as soon as we got out to the parking lot, we turned and it started caving in, and then it was over."

"At one point somebody ran up to the front door with an axe," Jody recalls. "My father said, 'I've got the key—you don't have to tear everything up.' Afterwards, the guy said, 'Mr. Sonny, you don't have to worry about that door now.'"

Marc Grollman remembers the immediate aftermath:

> *We all went to Jimmy Ewing's Circle Restaurant like we did a lot of times after a fire. Nobody believed Fisherman's burned down. Alo Klepper kept saying no way. We had to take a bunch of them there and show them.*

Sonny Schulz remembers, too. "A week later, we burned what was left to finish the job. We hauled out ninety truckloads of debris. That was a bad Christmas. We started rebuilding in February 1981 and opened back up five months later."

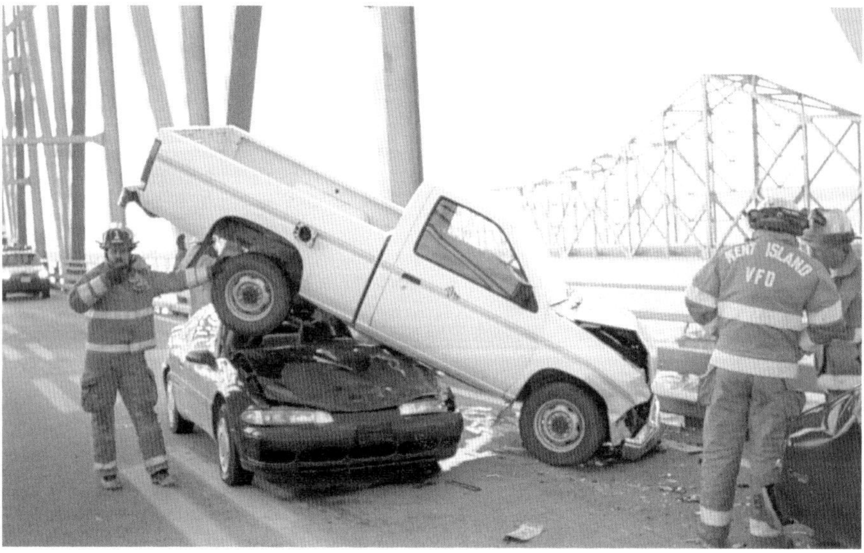

Only the guardrail kept the truck involved in this 1995 Bay Bridge accident from plunging one hundred feet into the bay. *Courtesy Tracy Schulz.*

The volunteers from the 1970s and '80s recalled the number of car accidents increasing around the time of high school graduation. Assistant chief Buddy Thomas said that New Year's often brought bad news, as well:

> *We had three or four rough New Years in a row. That New Years Day when Tommy Boyd, Leslie Griffith and those guys had that accident down Dominion, that was bad. People were lying everywhere. We knew them all, were in school with them. You looked, and they were on the side of the road, in the bushes. The ones who got thrown, it probably saved their lives. The car was wrapped around a tree.*

Marc Grollman talks about a later incident involving an island businessman and longtime KIVFD supporter:

> *When Mr. Robert Snyder died* [1987], *that was bad because he was so well liked. It was down Route 8, almost to Kent Island Estates. He was in his work truck, like he was 90 percent of the time, and somebody crossed the centerline and ran him off the road. His truck turned over, and he was killed. That was upsetting. Everybody knew him, and he was such a good friend of the fire company.*

Buddy Thomas says, "I'd worked with him that day and had just come up the road a few minutes before. I came up in the dump truck, and he was right behind me. I had one of those weird feelings something wasn't right."

"We had to send Billy Moore away," says Tracy Schulz. "He was too emotional."

Tracy, perhaps thinking about Billy Moore's own premature death, switches tone and topic. He laughs when he reminds everyone of Christmas 1997, when a camel escaped from the Kent Island United Methodist Church's live Nativity scene and ran down busy U.S. Routes 50/301. "A man called 911 and kept saying he thought he'd hit a camel. Ironically, he was from Saudi Arabia and talked with a thick Middle Eastern accent. The dispatcher asked him if he'd been drinking."

Emergency responders truly must be prepared for anything.

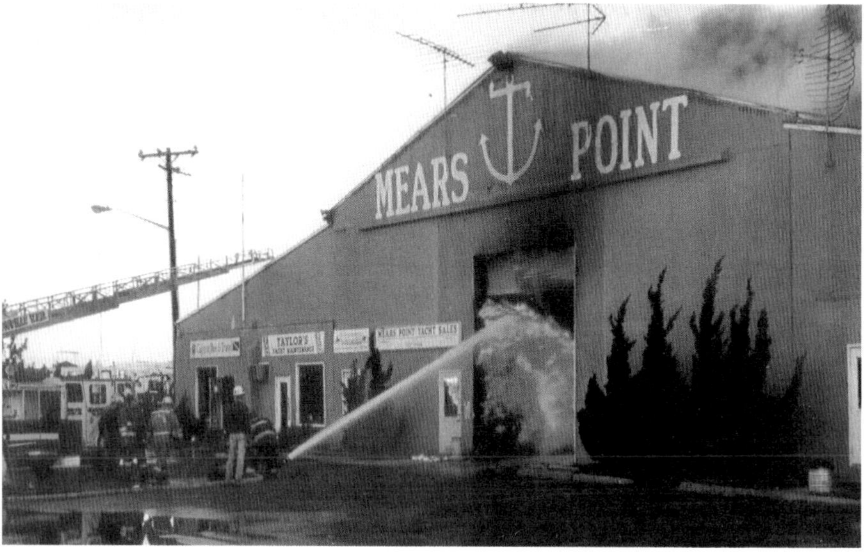

Faulty electrical systems sparked the 1998 Mears Point fire, a three-hour battle involving thirteen companies and one hundred firefighters. *Courtesy* Bay Times.

Modern Times

On September 11, 2001, members of the Kent Island Volunteer Fire Department erected a memorial in front of their Stevensville firehouse to honor victims of that day's terrorist attacks in New York; Washington, D.C.; and Pennsylvania. Within twenty-four hours, the display grew to include flags, flowers, candles, balloons and a boot for collecting donations.

The KIVFD also hosted a candlelight vigil. Maryland state trooper John Hall recited "The Final Inspection," a police prayer. Southern Kent Island's United Communities Volunteer EMS chief Mary Alice Van Hoy read prayers for emergency medical technicians and paramedics. Jeff Martin tearfully led the fireman's prayer: "When I am called to duty, God, whenever flames may rage / Give me the strength to save some life, whatever be its age."

While first responders across the country mourned the loss of their fellow Americans, and particularly their emergency services brothers and sisters, in the days after the 9/11 attacks, members of the Kent Island Volunteer Fire Department and the community they serve were soon to suffer a personal tragedy.

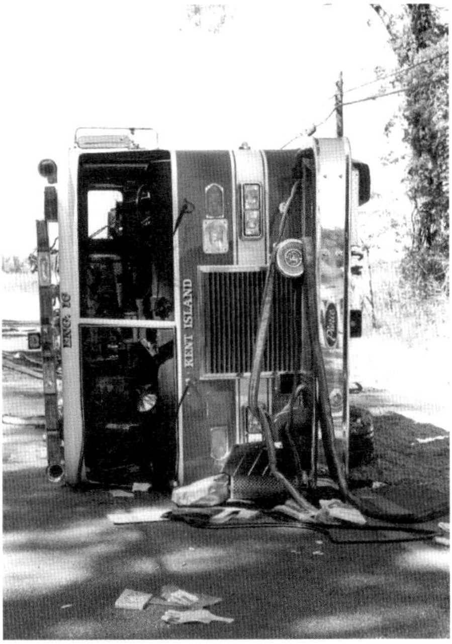

The death of loyal KIVFD member Dougie Thomas after this 2001 accident resulted in the department's increased emphasis on safety and training. *Courtesy Thomas family.*

Active life member and eleven-year KIVFD president Lou Johnson recalls:

> *It was the Sunday morning after 9/11, about 10:45 a.m., and second assistant chief Dougie Thomas was going to what was reported as a boat fire at the yacht club. When he went around that corner, the truck slid off the side of the road.*

Don Lewis was inside Engine #16 that late summer morning:

> *I was riding in back. Jeff Cresanto was in the front passenger seat. Jeff Clark was in back with me. To this day I couldn't tell you what happened. We were going down the road and for some reason I put my mask on. When I put my tank on, I leaned back and put it in the holder and was just sitting there when we had the accident. When the truck finally stopped, I was stuck. My arm was stuck—the bottle held, so I hadn't fallen anywhere. I was in there crooked and couldn't get out. I was hollering to Jeff. We were conscious. The pole just about took the whole side of the truck off. Luckily, the water tank stayed in place. If it had come loose, it would have crushed us. The truck was on its side. Dougie was laying there in the truck. He'd been thrown into the passenger side. He said, "I can't move." Everything*

happened quickly. Everybody was hollering, nobody was talking low, and they got help to us within seconds. Dougie kept saying, "It's getting harder for me to breath." I had struck the back of my head on that bottle. They flew Dougie and the other two to Baltimore shock trauma. They loaded me in the ambulance and drove me to the hospital in Annapolis.

"The boat wasn't even on fire," says Tracy Schulz. "It wasn't cooling properly and was steaming. Over the radio, we heard people yelling. When we got to the accident, everything was everywhere."

Lou Johnson was on the way to a Kent Narrows restaurant with his wife, Ronnie, a loyal ladies auxiliary member, to meet family for brunch.

We were the first car to come up on Dougie's accident. The truck was lying on its side. Hoses were all over the place. A couple of the guys were climbing out. Jeff Cresanto was up front in the officer seat, and Dougie was in a heap. When Jeff Clark got out, I went to Dougie, and he asked, "Are they all alright?"

That day, everybody came to the firehouse. The ladies cooked. In some ways, it felt like the old days. "When there's a crisis, you come together," says Don Lewis.

The accident left Dougie with a broken neck, paralysis from his chest down and without full use of his lungs. Doug Leatherman, a current member of KIVFD Board of Directors and a contemporary of Dougie Thomas, saw how the accident brought out the best not only in their organization but also in other area departments.

After Dougie's accident, he and Renee's house needed extensive renovations to accommodate his situation, and there was tremendous turnout from other fire companies to make that happen. It was great to see.

Jody Schulz notes, "It was almost a couple years from the time of the accident until Dougie died. He'd been doing better, and then his kidneys started failing. He fought hard."

Strong-willed to the end, Dougie Thomas died on July 9, 2003. Dolores Timms speaks about him with emotion:

There was a lot of trauma losing Dougie, all that time between the accident and the service. It brings home what the fireman's lot in life is. Every time you go out it's dangerous. My husband, Bobby, and I had just waved to that truck. We'd gone around the other side of the circle at Castle Marina Road when they passed. Dougie nodded. It wasn't fifteen minutes later my brother called, "Do you know your fire truck just had an accident?"

We take life for granted. When my monitor goes off, I ask the Lord to take care of whoever answers the call.

Link George concludes:

For Dougie's memorial service, we equipped the tanker so his casket could fit in it, and I had the rescue truck with the pallbearers. I'll never forget that day. I still get emotional sometimes thinking about it.

Hurricane Isabel blew into town on September 18, 2003, as a Category 2 tropical cyclone and remained at that highest level for almost two days, distinguishing Isabel as one of the longest such storms on record. Gusts up to 134 miles per hour were measured. Isabel pushed so much water so powerfully into the Chesapeake Bay that normal tidal cycles were interrupted. Scores of Kent Island homes and businesses were damaged by the record-breaking tidal surge.

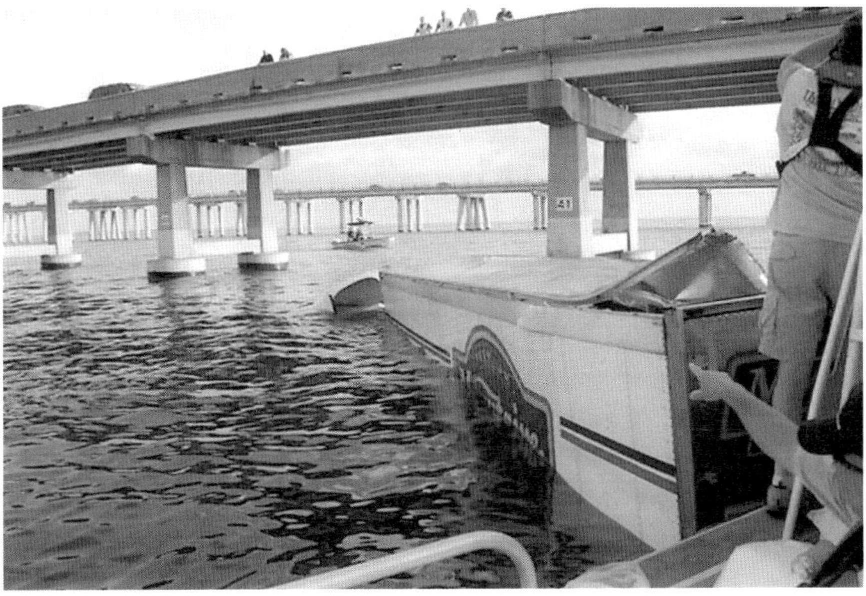

In August 2008, a fatal early morning accident on the Bay Bridge left a tractor-trailer partially submerged in the Chesapeake. *Courtesy Will Clark.*

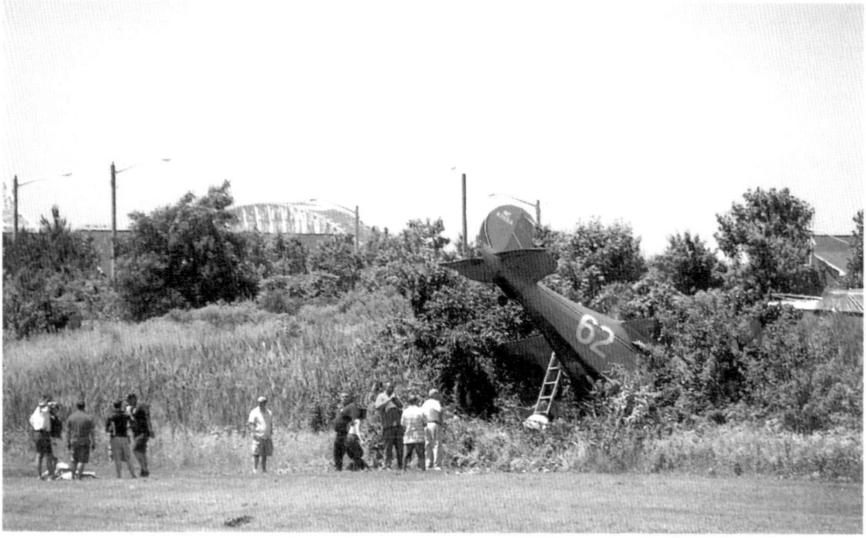

In June 2009, first responders arrived at Bay Bridge Airport to find a crashed vintage airplane. No one was seriously injured. *Courtesy Will Clark.*

Superintendent of machinery Tom Balderson describes the experience from the perspective of the fire department:

> *We made fifty-seven calls in twenty-four hours. Mostly going to get people out of places they shouldn't have been. They were told to leave. There's a six-foot tidal surge coming. There's a good chance your house is going to get it. Few hours later, they've got a foot of water in their living room; they call and say, "We're ready to go now."*

Jody Schulz touches on what had changed since he joined the department in 1978. "Expectations have changed. In the old days, people were happy to have a fire truck show up. Now they want to know what took you so long."

Former firefighter and retired high school math teacher Jay Gibson understands, though. "When you're the one waiting, even a little seems like an awfully long time."

Paul Fuld, the oldest active member and the man known to everybody as "Friend," laments the destruction he has witnessed. "It's horrible when someone loses their life, but I also hate to see people lose their property. It's sad to see people lose everything they own."

Curtie Chance says, "First time you see somebody die in a fire, that's something you don't forget."

Longtime friend of the KIVFD Bobby Norris agrees:

> *One of the worst was when Turkey Baxter's son burned in a tenant house fire at Grollman's Farm near Love Point. There was nothing we could do, and he was just a little boy. That was a real bad day. You never get over it.*

Doug Leatherman talks frankly:

> *I was an EMT for seventeen years, but I've gotten away from it. I had two children die in an automobile accident, and I started drifting away afterwards. When I look back on why, and if I'm honest with myself, that's why I stopped going on the ambulance.*

Calvin Duncan adds:

> *To this day, I've got a little girl who stays in my mind. She was only six months old, and I went to Harbor View one night to pick this baby up. They'd just come back from Baltimore and thought it was carbon monoxide from the exhaust. She was on the backseat. We went in, and they couldn't find a pulse. Duke was with me. I said let me check and I found a pulse. We get in the ambulance, and I gave her mouth to mouth all the way to the Annapolis hospital. They took Duke and me back into a room. The doctor came in and said, "That baby was dead when you picked her up." I said, "No way. She couldn't have been. I felt a pulse." He said, "No, you wanted to feel a pulse so bad you felt your own pulse through your fingertips."*

"When you deal with children," Calvin says quietly, "it's terrible." No one elaborates.

Part V
BRIGHT AND HAPPY EXPECTATIONS

If you paint in your mind a picture of bright and happy expectations, you put yourself into a condition conducive to your goal.
—Norman Vincent Peale, American preacher and author

The Carnival

If your goal is to help people in their most desperate moments, there's a need to witness a few bright and happy expectations. The *Bay Times*, on July 6, 1988, reported:

> *When multi-colored neon lights are flashing, the music is blaring, daredevils line up for thriller rides, contestants gather around game booths and the sweet aroma of the pink colored cotton candy, corn dog, and popcorn are in the air, you know you're at a carnival.*

The carnival is important as both a fundraiser and a community event. It brings the KIVFD membership together with the public they serve in a celebratory manner and has since the beginning. On August 1, 1947, at a special meeting of the Kent Island Lion's Club, fire department and ladies auxiliary, Mr. Walter Denny was elected chairman of the committee responsible for the first carnival.

His committee worked fast. The carnival, held on the grounds of the Stevensville High School, started on Monday, August 25, and continued

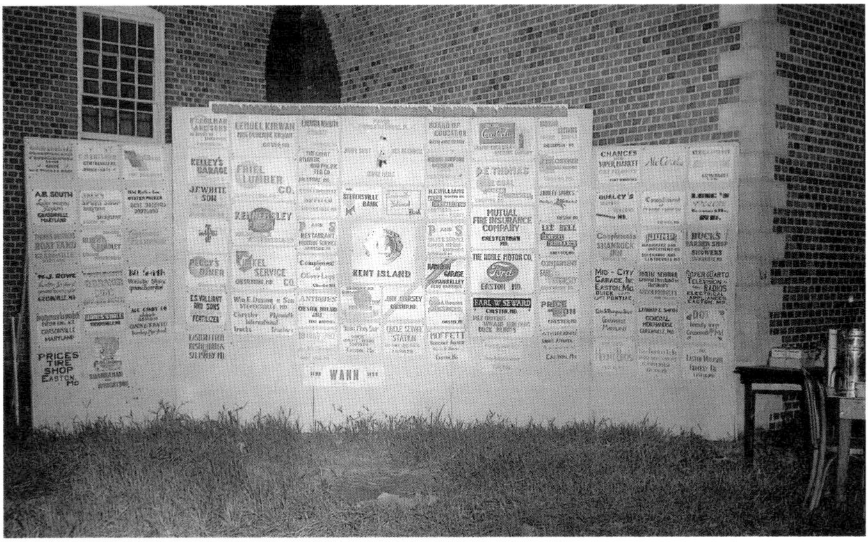

Jim Ball's hand-painted advertising signs were traditional carnival fundraisers. Each colorful sign was unique. Together, they created a huge folk art display. *Courtesy Kirwan Collection.*

through Saturday, August 30. A concert by the Eastern Shore Playboys and a musical film presented in something billed as "Technicolor of the South," were two of the week's featured events.

Prizes—items donated by various individuals and businesses—were raffled off every night. Lee Bell's Sales & Service in Chester contributed a GE automatic iron and a radio. Chester Holland from Grasonville offered an antique vase. Chester's Michael Pelczar donated a canary and cage.

Out-of-towners ponied up, too. The Katz brothers, jewelers from Baltimore, contributed two lapel pins. Lever Bros. Co., from Cambridge, Massachusetts, donated a case of its popular laundry soap powder, Rinso. The Grand Ground Prize, a pig donated by Chester's Mr. and Mrs. John C. White, was drawn on Saturday night. Dolores Timms acknowledges, "Thousands of raffle tickets have been sold over the years for everything from a carton of cigarettes to a Valiant automobile. They made a lot of money on raffles."

Billy Denny says, "In later years, my dad would give them a car at cost to raffle off. I think chances were five for a dollar."

From the beginning, the carnival was a financial success. The *Queenstown News* announced on September 1, 1947, that "thanks to the support of the entire county, the Kent Island Volunteer Fire Company went over the top!" The net proceeds for the first carnival reached an astounding total of $3,175.48.

The tightknit but not overly economically powerful community made sure to patronize the KIVFD's major annual event. Kitty Duncan puts it this way: "You never had any extra money all year but come carnival week you'd have money every single night." The carnival was that kind of tradition by its second year. The *Queenstown News* of August 13, 1948, reported:

> *The biggest and best Carnival of the Eastern Shore Summer Season will be held on the grounds of Stevensville High School from August 23 to August 28. Sponsored by the Kent Island Lions Club for the benefit of the Kent Island Volunteer Fire Department, the Kent Island Carnival is a real, all-out good time affair.*
>
> *As Carnival week approaches, every man, woman and child resident of The Island goes into action to make this another rip-roaring success. Everyone will be working, from committee chairman smoothing out last minute details, housewives baking cakes, men erecting stands and lighting, teenagers help in numerous ways, down to little tots hoarding their pennies for rides on the famous Choo-Choo Express, the cheapest ride on the Eastern Shore at nine cents, three for a quarter.*

Things didn't always go as planned, of course. In 1948, the famous radio entertainer Happy Johnny was scheduled to appear when rumors that he died began to circulate. Some testy correspondence resulted between fire department representatives and the company contracted to provide entertainment.

Secretary Virginia Bell wrote to David Kay of Philadelphia's Kay Amusement Company on July 23:

> *Since our conversation this morning I have been informed by a number of people that Happy Johnny is dead. Please advise who is running the show. Does it have the drawing power that it had prior to his death? Inasmuch as I am too busy to listen to daytime programs I have lost track of the program in question. I shall have to leave it to your discretion as whether or not the program minus the original Happy Johnny will pull in the crowd sufficiently to warrant our having him.*

Mrs. Bell asked for a quick response since she anticipated being "swamped with questions and criticism." David Kay responded three days later:

If you and the people who are bothering you would like to see a "live Corpse," come out to Elkton, Maryland this coming Thursday night and see Happy Johnny in "the flesh," where he will be entertaining for the carnival.

I don't know where this rumor started but that is just what it is, only a "rumor." Please leave the promotion and entertainment up to me. I have been in this business for years and have so far taken care of all the details successfully, and won't let you down either, I promise.

Mr. Kay went on to explain that he was having trouble finding the bumper cars the Kent Islanders wanted and that Happy Johnny was a rain-or-shine event. In conclusion, he stated that a Happy Johnny show one rainy night in nearby Denton grossed $1,600, and Mrs. Bell could "check on this if you so desire."

No one remembers for certain, but it's assumed that the Happy Johnny controversy worked out satisfactorily in the end.

That problem was taken care of behind the scenes. Some embarrassments were more public. Quite a few islanders remember when Maryland governor William Preston Lane Jr. was a special guest, and the chauffeur drove his

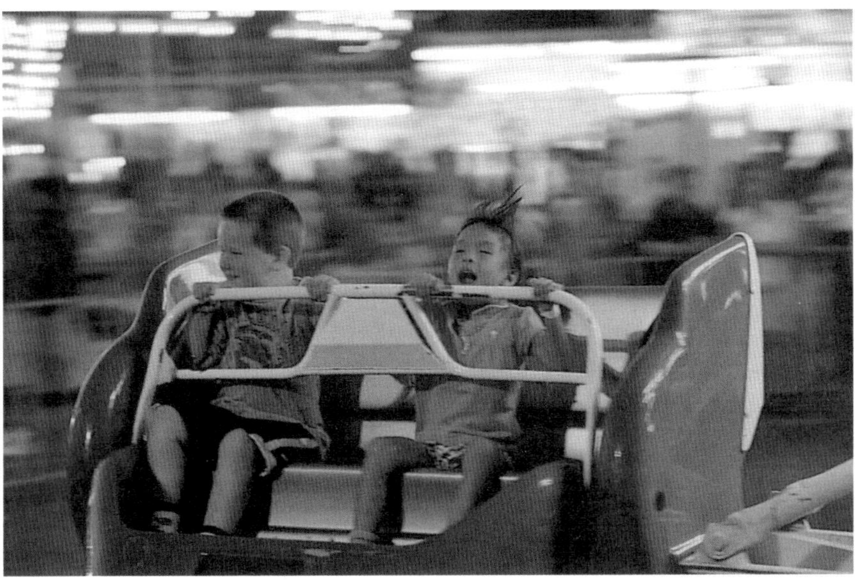

Whether it's the Scrambler, Ferris wheel or merry-go-round, the annual Kent Island Fireman's Carnival provides amusement rides for diverse thrill levels. *Courtesy Will Clark.*

limousine right into a huge ditch in front of the old high school. Bobby Norris grins. "It was the talk of the island for several days."

Those early carnivals combined an opportunity to experience thrills, both personal and public. Babe Grollman says:

> *One time they had airplane rides, and I bought a winning ticket. We circled Kent Island. Wheat piled in the fields for thrashing looked like sand piles. People looked like ants. There weren't many cars at all.*

Billy Denny, on the other hand, remembers, "One time Jimmy Lane wrestled a bear, a great big black bear. They sold tickets to watch, and had a ring set up and everything." He shakes his head and laughs. "Old bear beat him pretty good."

Dolores Timms recalls, "There was a talent show, and they'd crown Miss Queen Anne's County at the carnival."

Ray Baker, now seventy-seven years old, and his father both spent decades of time and effort helping oversee successful carnivals:

> *As the years went on, we learned the best way to have a successful carnival was to get kids to come. After a while, I became entertainment chairman. I had a budget of $300. We'd have skydivers, Baltimore TV stars like Pete the Pirate and big-name ballplayers.*

One of those ballplayers was a popular all-star third baseman for the Baltimore Orioles. Ray continues:

> *Brooks Robinson was coming one year, but this big summer storm rolled in. I called his home, and his wife said he was out but would call me when he got in. I had all these people excited to see him, but it wasn't going to happen. I told my daughter I was expecting his call. When the phone rang, she answered it and handed it to me whispering, "It's him, it's him." We talked and decided to call off the appearance because we had no inside place to put him. He was very understanding. When I asked him what we owed him, he said, "Nothing. I didn't come." He was very nice about it.*

Though entertainment and rides were in some degree or another hired out, food concessions were provided by members of the fire department and their support network. "In the early years," says Billy Denny,

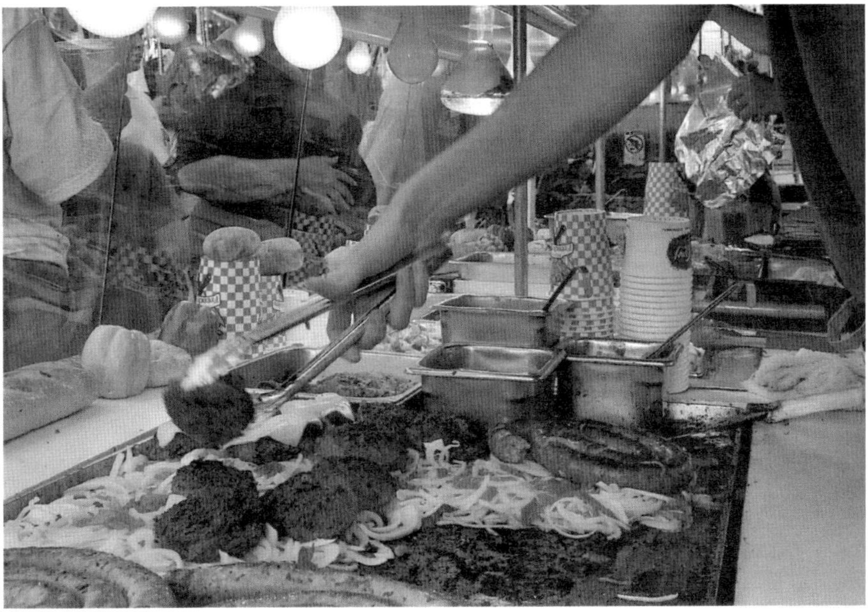

Grilled sausages, cheeseburgers, pit beef, fried clams, hot dogs, ice cream and cotton candy—nothing tastes better than carnival food. *Courtesy Will Clark.*

> *it was a real community effort. Farmers donated chickens and vegetables. Soda companies donated cases of soft drinks. We'd get hamburgers and hotdogs from neighborhood stores. I'm sure Mr. Frampton's store in Stevensville gave a lot of meat away over all those years.*

Yet there were costs. The food booth expenses for the week of August 28 to September 2, 1950, totaled $60.14. That grand sum included fifteen pounds of crabmeat, forty-two pounds of ground beef, forty-five pounds of onions and six pounds of lard.

Charter member Billy Lane was a loyal carnival week volunteer:

> *For a long time I was head cook and manager of the hamburger booth. Mr. Pat Thomas* [Percy "Pat" Thomas, 1900–1976, native Kent Island storekeeper and civic leader] *ordered the meat and bread through his store. We'd all help each other. Wednesday night was the parade and our biggest night. We'd start peeling onions that afternoon. My mother, father, wife, everybody in the family helped out.*

Bruce Coursey is among those who affectionately remember the dependable presence of Helen May Watkins and her brother, Jimmy. The Watkinses were born and raised on Kent Island's Cox Neck Road. For almost thirty years, Helen and Jimmy worked the carnival food booth and assisted in the kitchen for such firehouse functions as installation dinners. Jimmy is a soft-spoken and friendly gentleman. Helen used to be a big woman with a personality to match. She ran things her way. Bruce called her the "Hamburger Queen." Though she was funny and sociable, the queen took no guff. Health problems have reduced her weight, and she misses her fire department friends, but at seventy-eight, Helen's still a force to be reckoned with.

Jimmy recalls the carnival's genial neighborly atmosphere. "You could always feel the love," he says.

Ray Baker witnessed the birth of a new culinary trend:

> *We started frying clams. A man from the Maryland Department of Natural Resources was promoting them. It wasn't anything locals ate much before. Watermen used them for bait. We got cookers. The clams were delicious. They really took off. Later on it became a big item at area seafood festivals.*

To the present day, KIVFD provides as much of the food concession as possible. Member Todd Kerchner says:

> *For the last few years, I've worked the pit beef booth. We do five or six twenty-pound pieces of beef every night. I look at that meat when we get started and say we'll never sell all that. Sure enough, we always sell it.*

When the carnival was held in the parking lot of the old Stevensville High School, across the street from the firehouse, the first step in setting up was unpacking the booths and other structures. The second was dragging them across the street. Jay Gibson was always part of that work crew and knew it was no easy feat. "We did it all by ourselves. Everything was owned by the fire department except the rides." Calvin Duncan believes there was a time when even a couple of the kiddie rides were owned by KIVFD.

The ladies auxiliary membership is a big component in the carnival's success. Attendees could always look forward to seeing the smiling faces of the lady volunteers working everything from the Goldfish Booth to the Pick-up Ducks Stand. Cindy Taylor grew up in a family devoted to the KIVFD; her father, Ellwood Smith, was a life member. Cindy fondly mentions one of her favorites—"the booth with the dishes. You'd try to toss a nickel into one to win."

Shirley Thomas remembers, too. "I worked that booth many nights, and the one with the wooden bottles and stuffed animals you'd throw rings over."

The Cake Booth is one of everybody's favorites. Thousands of delicious homemade cakes have been won over the years by folks betting on the spin of a numbered wheel. That means thousands of cakes needed baking. The ladies made sure that happened, and they still do. Kitty Duncan says, "My cousin from Philadelphia comes down every carnival week to bake cakes and has been doing that for over thirty years."

"It was always fun," says Dolores Timms. "Everyday a cake had to get broken. We'd eat it for lunch." She giggles when she thinks about it. "And every day the kids walked the carnival grounds to find coins dropped the night before. That all went to the firehouse. The carnival was a big thing. A big thing for families, a big thing for all of us."

On occasion, a recent loss gives carnival week a different vibe. On August 26, 1970, *Delaware State News* reporter and one of Queen Anne's County's most prolific and well-loved journalists Dan Tabler wrote:

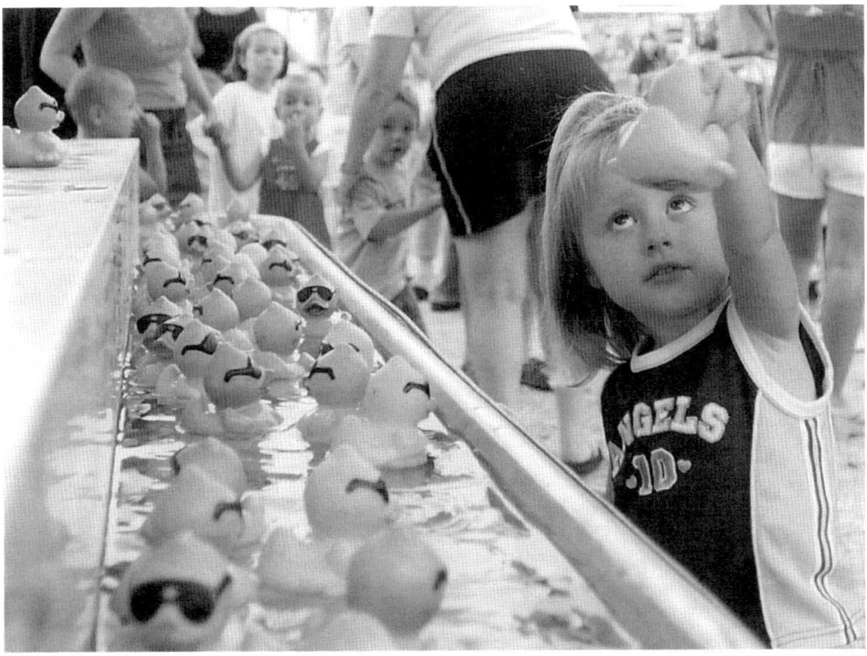

Pick-up Ducks rate with the Cake Wheel and Goldfish Booth as perennial carnival favorites. *Courtesy Alison Harbaugh/www.frecklephotography.com.*

> *This year a pall has been cast over the usual gaiety, the bright lights, the thrill rides, the parade itself, and the many booths that make up the midway at the Carnival. Chief John Holden is not here to lend a hand. John died just two weeks ago of a heart attack, at the all-too-young age of 50.*

Marc Grollman remembers carnivals of the 1970s and '80s:

> *The carnival was held behind the firehouse when I was active. There was a big dinner the night before the carnival started. The food was out of this world. Most of us did whatever needed to be done that week. Clint Phillips and Paul Fuld worked the information booth and called ground prizes. Duke sold tickets and oversaw things. Cleanup was a big day. Sometimes it took more than one. But it didn't matter if you were sick or exhausted; you showed up to help.*

Kenny Bullen, another volunteer whose family was passionate about their KIVFD service, thinks back to when convicts housed at the old state

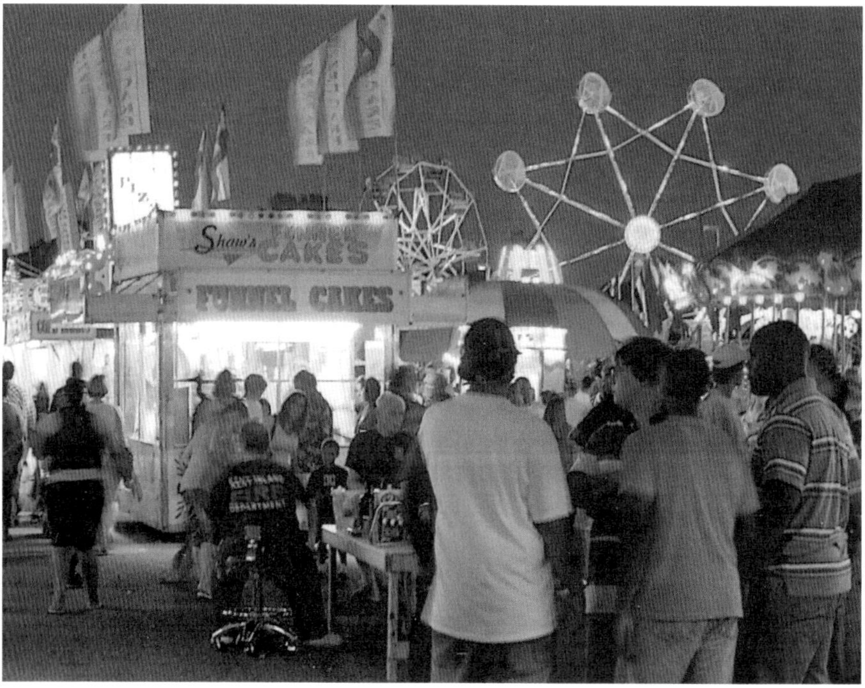

Carnival week remains a highly anticipated KIVFD tradition. *Courtesy Will Clark.*

prison camp on Cox Creek would help dismantle the carnival grounds. "You couldn't set it up in a day," Kenny notes, "but with enough help you could take it down in a day."

The Kent Island Volunteer Fire Department's annual carnival has been a summer benchmark for generations. Today's carnivals are expected to bring in between $40,000 and $50,000 in funds. Many of the members give credit for the carnival's continued success to current vice-president Don Lewis, and the KIVFD is grateful that Queen Anne's County Board of Education allows the carnival to be held on the grounds of the Kent Island High School, which opened in 1998. Staging such a large event requires even more from the volunteers than they already give, and today it is far more complicated than it once was. Life member Pops Gellert thinks of all the work putting on a successful carnival requires. "Carnival now takes a month before and two weeks after," he says, "but people only see one week."

Bobby Norris, however, considers the carnival from a different perspective. "The carnival was great. It was where you either got a girlfriend or you lost one."

Kent Islanders tend to be romantics.

Parades

The Kent Island Fireman's Parade follows a long tradition in America's fire service. From the earliest days, firemen from competing companies dressed in their finest regalia, polished their equipment to perfection and marched up some town's Main Street to cheering crowds. Bobby Norris expresses how important parade participation used to be:

> *There wasn't anything bigger than the annual fireman's carnival, and parade night was really special. The first time I got to drive one of the fire trucks in the parade, they let me drive the second truck purchased, the 1951 International pumper. I was bursting at the seams with pride, and had butterflies in my stomach the entire route.*

In the beginning, there were usually two parades: the official Fireman's Parade, with companies from all over the region competing for awards; and the Clown's Parade, or Variety Parade, where groups, organizations and prominent individuals dressed in outlandish costumes. The crowds loved it. Participants maybe loved it even more.

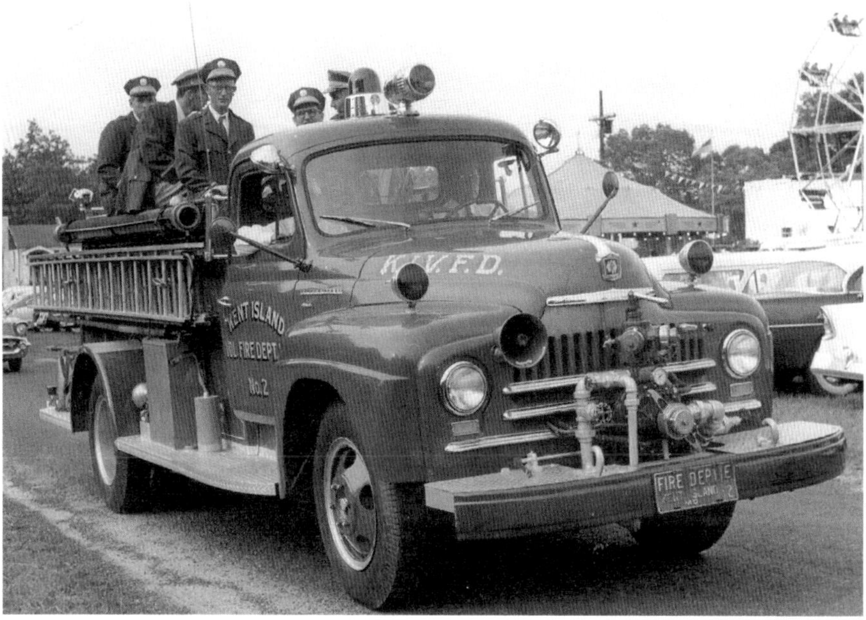

The Fireman's Parade was a highlight of Kent Island summers for decades. Island firefighters were equally excited to participate in other communities' parades. *Courtesy Joyce Moore.*

Shirley Thomas remembers both parades:

> *We'd go over Francis Price's and sit on his fence and watch the Fireman's Parade for hours and hours. The parades used to be huge. Fire departments from all over came, marching bands. It was really something to see. The Variety Parade was great, too.*

Dolores Timms admits, "We had as much fun preparing for the Variety Parade as we did the actual parade."

Cindy Taylor says, "Jackie Moore and my mother, Leona Smith, their float was always a secret. As kids we'd sneak and try to see what they were doing, but oh, no, that wasn't happening."

As much fun as the Variety Parades were for all concerned, participation dropped off over time, and they eventually ceased to be part of the festivities. The Fireman's Parade, however, continued to grow in size and difficulty to organize.

The August 29, 1968 issue of the *Bay Times* reported:

> *A crowd of over 3,000 people jammed Stevensville for the annual fireman's parade held during the K.I. Fireman's carnival that will continue thru Saturday. Hundreds and hundreds of bands, majorette groups, and pieces of fire equipment passed in review Tuesday night in a parade that ran for over 2½ hours. "The largest parade ever held in Queen Anne's County" was the general comment of bystanders.*

Of course, to get other fire departments to participate in the Kent Islanders' parades, the Kent Islanders were expected to reciprocate. Travel and the opportunity to stand out amongst peers were exciting propositions for the members of KIVFD, particularly when that opportunity came from "the Other Side" of the Chesapeake Bay. Ray Baker reminisces:

> *The only chance we ever really got to show our pride and puff our chests a little was at a parade, especially on the Western Shore. We'd be all dressed up. Donald Thompson and I used to ride in Number 11. We liked riding in that truck. I remember riding to Powhatan Beach and other places I'd never heard of. People sitting on the side of the street would look at us, then look at each other and ask, "Kent Island? Kent Island? Where's Kent Island?" Somebody else would answer, "That's across the bridge." And they'd say, "Way over there?" We'd tell them, "It's really not that far."*

Bobby Norris adds:

> *The night the ladies auxiliary won first place in the Glen Burnie Fireman's Parade was a super accomplishment in those days. I was calling cadence, and when we got in front of the judges stand we put on a show. We'd been practicing all week behind the old high school and were very confident. We were strutting our stuff. We spun around so many times I had to ask Miss Lottie Taylor which way we were heading.*

The parade tradition carried on through the 1970s and '80s. Marc Grollman recalls that in his era, KIVFD continued to participate in parades. "We went to parades from up past Baltimore to Delaware and everywhere in between. We often got trophies for Farthest Traveled." He is quick to add:

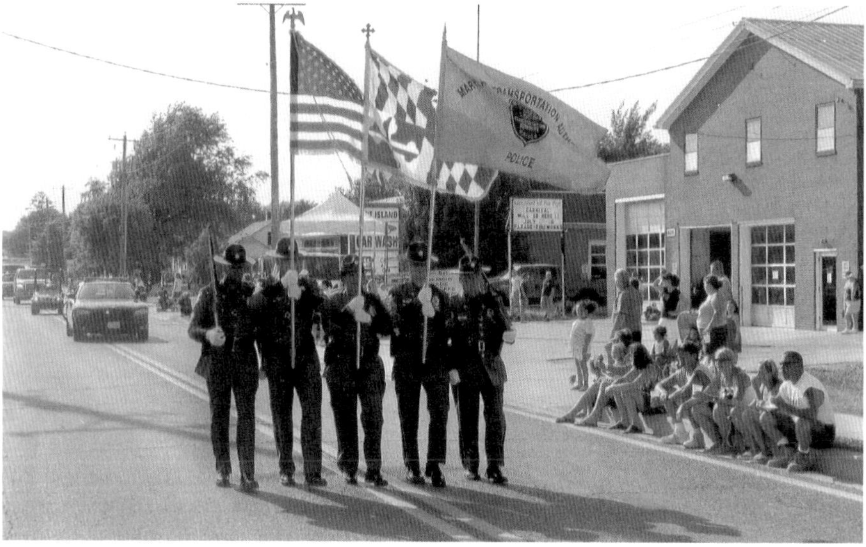

Finding organizations and entertainers to participate in the annual parade has become increasingly difficult. The 2009 parade was KIVFD's era-ending last. *Courtesy* Bay Times.

> *We'd get trophies for best appearing pumper or ambulance, too. There were some we'd go to every year like the St. Michael's Firemen's Parade. We started going to Ocean City every year for the Maryland State Fireman's Association parade.*

In more recent years, however, it has become obvious that the huge hometown fireman's parade is a thing of the past. Fewer communities host parades, so a shared commitment to participate has waned. Expense and time constraints top the list of challenges out-of-town fire departments face when deciding whether to attend. Marching bands willing to compete for trophies are scarce in the twenty-first century. Entertainers want to be paid. Driver Pops Gellert says:

> *When I first started driving in 1989 or '90, I went to twelve, fifteen parades, easy, a year. Now we might go to eight or ten at most. Depending on how close the parade is, we take one to three trucks. But in general, parades are getting smaller and smaller. Used to take us forty-five minutes to run the parade; now it's about twenty. Engine 11, back when Calvin was chief, we'd spend two days shining it for a parade. We'd take a trophy every time we went somewhere. It was fun hosting parades, too. Those days are over.*

Because of extreme difficulties in staging a parade in the modern era, in May 2010, members of the Kent Island Volunteer Fire Department voted that the previous year's Fireman's Parade would be their last.

Fundraising and Social Events

Fundraising started before there was much of anything else. In May 1947, Catherine Kirwan, the senator's daughter, civic booster and ladies auxiliary organizer, sent out the following notice:

> *A dance will be held each Saturday for the benefit of the Fire Company. Mr. Ball and two members of the Ladies Auxiliary will be in charge. Cokes and sandwiches will be sold. A dance will be held at the Island Casino Saturday, May 31, 1947.*

Billy Denny recalls:

> *Kent Island Casino was where Ram's Head Shore House is now. Elizabeth Potts owned that. It was a bar and had a big dance floor. You went there to dance with girls or watch wrestling on one of the first TVs I remember being around here.*

Department historian Dolores Timms displays a china set:

> *Mr. Billy Lane told me these dishes were used for our first dinner. Records show the first dinner was held in April 1947. The firehouse wasn't built until 1948. They held these dinners at what they called the Community House, which is where Billy Denny has his antique shop. That was the first church hall and belonged to the Methodist Church.*

Billy Denny confirms, "That building was a church turned into a community hall. Church, the fire company, Lions Club, Boy Scouts—whoever needed it, could use it."

After the firehouse was built, the upstairs banquet room became the place to hold certain functions. It was the biggest and most modern facility available on Kent Island.

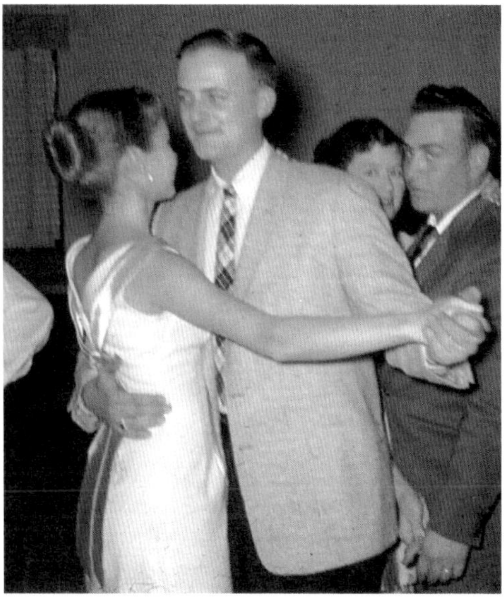

Blind date. Bobby Norris met Genevieve Sparks at a February 1959 installation ceremony. The couple celebrated their fiftieth wedding anniversary in June 2009. *Courtesy Bobby Norris.*

Dolores Timms lists some of the functions the firehouse hall hosted:

> *Dinners, dances, New Years celebrations, election day dinners, fireman's association dinners, breakfasts, wedding receptions, birthday parties, visits from Santa Claus, weeknight bingo, Sunday bingos and all-day bingos. Bingo has been a great fundraiser from day one. The first bingo was noted in the 1947 minutes. In 1947, they organized the first dinner, the first carnival— before they had a firehouse or a piece of equipment. The women were always working hard, always fundraising.*

Kitty Duncan memorializes their longtime bingo caller Mr. Fonzy Thomas:

> *He'd crack you with his cane if you didn't steer clear. But he was faithful to us; he'd sell chances for anything for us. Right to the end he was famous for selling fifty-fifty raffles. His wife Catherine was in the ladies auxiliary.*

Dolores mentions the auxiliary's continually successful bingo operation. "Sunday's a good day for our bingo. We have our regular players, but there are a lot of new faces. The new facility attracts people."

In the 1980s, the KIVFD hosted regular adult dances, fitting the image of the brave firefighters blowing off steam—and lots of it. Known to enjoy a laugh, Lou Johnson shakes his head:

> *Those dances were really something. Bring your own beer. Guys were carrying washtubs full of beer and quarter kegs up the stairs. I thought, What ever happened to six-packs? Music would be going, and the floor beams would be jumping up and down, bending. Curtains moving with the music. One of the last parties up there, Billy Moore got this band, said we had to have them, they're great. They must have had twelve pieces; they were halfway down the floor. Must have been 150 of us crammed into the last fifteen feet of floor space.*

"A dance wasn't over until Bruce Coursey swung from the rafters," Jody Schulz laughs.

"And did his striptease," Bobby Timms adds, laughing even harder.

After Kent Island's McDonald's opened in 1985, the fast-food restaurant invited KIVFD members to cook breakfast one morning. Proceeds from sales of certain items would be donated to the fire company. Lou Johnson says:

> *Must have been twenty of us there; they had room for two. They had so much help they didn't know what to do with us. We ended up cooking, washing windows, cleaning the parking lot. I think we had a pancake-eating contest.*

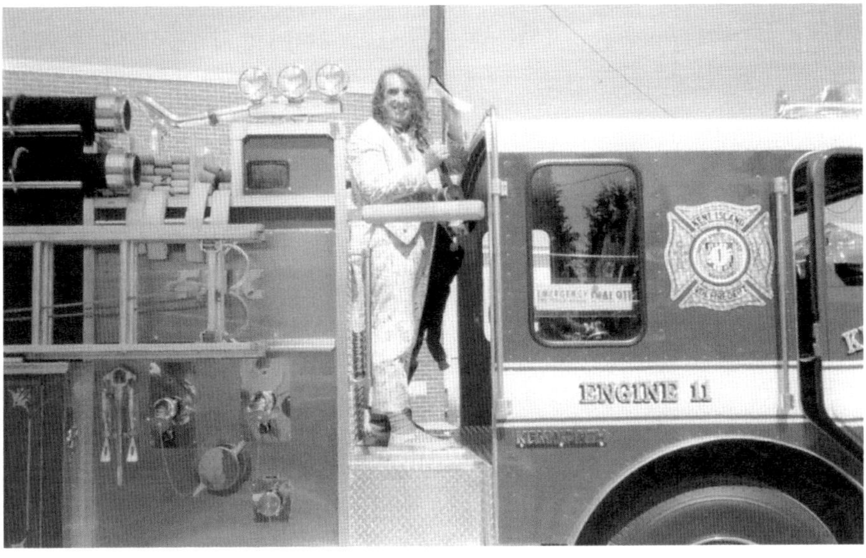

Traveling with a small circus in 1985, novelty singer Tiny Tim was treated to a fire truck tour of Kent Island. *Courtesy Joyce Moore.*

Calvin Duncan notes, "I think we made a grand total of $170."

"They were glad to get rid of us," Pops Gellert reminds everybody.

The fiftieth anniversary of the KIVFD was in 1997. The February 26 edition of the *Bay Times* reported:

> *Fifty years of hard work, dedication, and brother and sisterhood were recalled Saturday night as the Kent Islander Volunteer Fire Department celebrated its golden anniversary at a gala extravaganza. "I've really been nervous about this," said Dolores Timms, longtime member of the department's ladies auxiliary. "I've never spoken to such a large crowd and I'm not sure if I am qualified to speak." But Timms was more than qualified: her recollections brought many smiles, and a few tears, to the crowd.*

In 2003, a new fundraising idea was announced. The *Bay Times* described the endeavor on September 10, 2003:

> *The Gents of Kent Island want to spend 2004 hanging around your kitchen or in the laundry room. Or perhaps even in your bedroom. The KIVFD has come up with a unique idea for raising money for their new firehouse: strip men of their clothes, take their photo, and put them on a calendar.*

Media-savvy volunteer Terry Smith has been involved with the calendars since the beginning. She contributes an insider's perspective to the proceedings:

> *All the shoots were hilarious. Usually, the participants knew each other. They'd be apprehensive at first; next thing you know, they're rolling their pant legs up, taking off their undershirts, making fun of each other's knobby knees and pot bellies. Butterball* [undefeated local anchor-throwing champion Calvert Thompson] *is in every calendar, and he's a lot of fun. One year, right there in the Narrows, in front of everybody, including his work crew, he stripped to his white waterman boots and a pair of Spongebob Squarepants Christmas underwear.*
>
> *The insurance agents were fun, the restaurant owners and chefs, the farmers at Claude Lowery's farm. When they pull out that antique tractor and you see people like Paul Gunther and Babe Grollman enjoying themselves so much, it's great. We've shot on docks, in doctor's offices and grocery stores. When we shot the Red Hat Ladies at Lowe's Nursery, the women looked like they might really be nude because of the high flowers and shrubbery. One customer said he planned on staying all day. The cake*

ladies' shoot was a family effort. My mom was there, my sister-in-law, Cindy Taylor, and my niece. Shooting the firemen though, that's my favorite. Those guys are so much fun. I love them. They're all into it and have such great ideas. People start lining up on the street to watch. Anytime you see a bunch of naked firemen, it gets your attention.

The calendars became popular collector's items. Some high-profile individuals appeared in various editions, including former governor William Donald Schaffer, legendary Baltimore Orioles first baseman Boog Powell and longtime Baltimore Ravens football coach Brian Billick. Large signing events were held at various venues over the years, and the photo subjects became celebrities themselves. Pops Gellert's newfound pinup fame brought him some embarrassment. "I walked into my daughter's house, and her mother-in-law was there. She said, 'I didn't recognize you with your clothes on.'"

Unfortunately, the annual calendar project began to lose steam. The calendar's last year of publication was 2009.

In 2003, KIVFD participated in another unique and high-profile fundraising effort when the Kent Island Volunteer Fire Department's entry

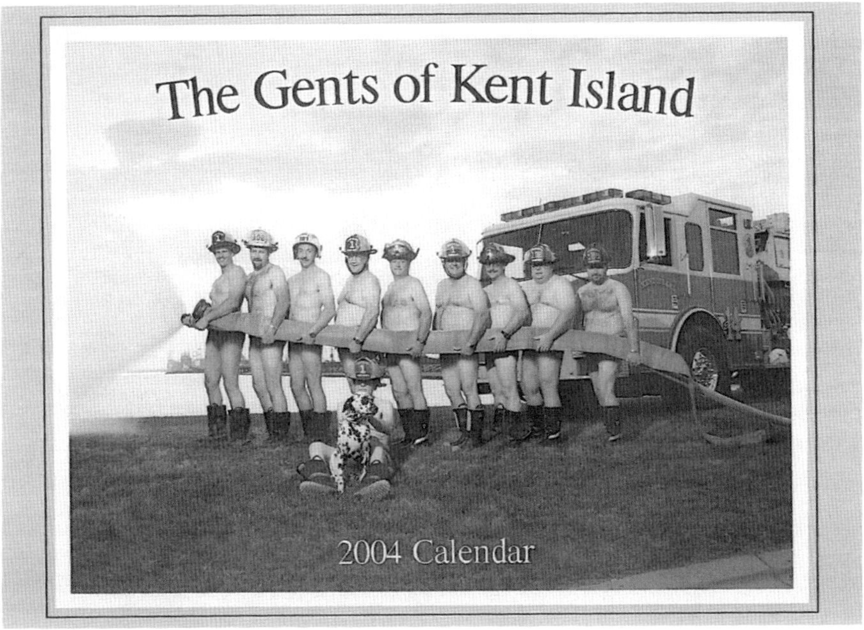

For several years, KIVFD's "Gents of Kent Island" calendars were a popular fundraiser that captured the attention of the community. *Courtesy AsiPhotos.com/Stan Ruddie.*

was named one of fifteen finalists in "The View from Your Hometown" contest on the ABC television show, *The View*.

Once again, Terry Smith was there:

> The View *had a contest—"The View from Your Hometown"—where local nonprofits could raise money for their organization. They required you send in a video of your nonprofit and why you should win. We only had a matter of days once we got started. We decided since everyone was going to submit pretty straightforward videos, we'd do something funny. We had Jody* [Schulz] *running through a field, looking like he didn't have anything on. We got* [sheriff deputy] *Walter Clough Jr. to pretend to arrest him and said we need the money because we had to keep bailing Jody out after calendar shoots. We also sent a bunch of calendars to the show. We came in second place but still got a nice cash prize.*

Dolores Timms credits much of the department's financial success to supportive civic organizations.

> *The Lion's Club was very important in the beginning. Today, the Kent Island American Legion is so good to us, gives us so much money, four or*

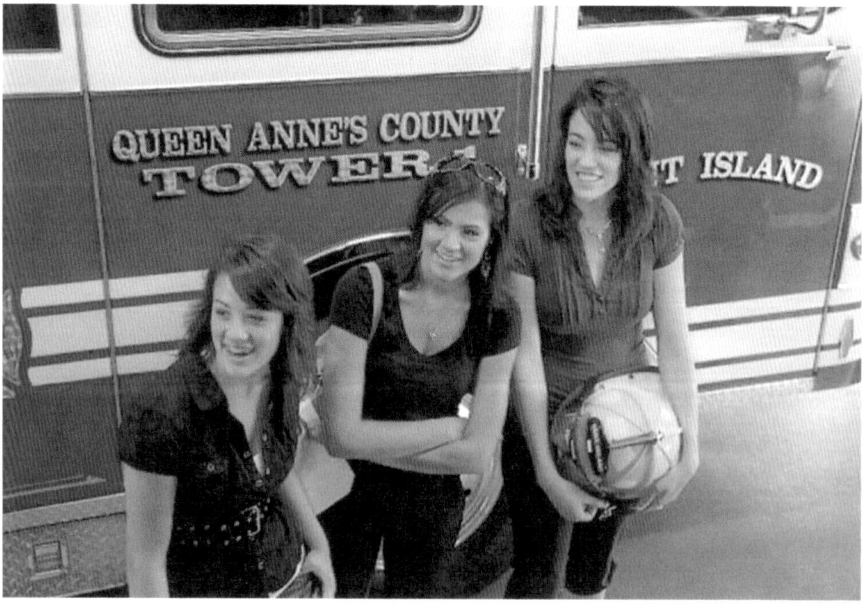

In 2009, the Kent Island–based band Beyond the Veil performed two concerts and produced a promotional video for the benefit of the KIVFD. *Courtesy Beyond the Veil.*

five times a year we bake cakes for their various functions. The fire company and the American Legion have worked together for a long, long time.

Life member Doug Leatherman concurs.

I'm not sure there's anything within reason we could ask the American Legion for they wouldn't give to us. There are other organizations, too. The VFW is good to us. The Moose is very supportive of fire services. In the business community, the Kent Island Advisory Council really helped with fundraising and such for construction of the new firehouse. We'd be lost without them.

Though money comes from many sources, securing grants is a priority. Jody Schulz says:

Grants have made a big difference in fundraising. They've helped get us turnout gear, air packs, all kinds of stuff, particularly after 9/11. Paul Schlotterbeck, our current chief, is good with grants. He's done a lot of hard work in that area.

A number of current members also give Jody credit for employing his successful business background to the fire department's financial advantage.

Terry Smith talks about recent developments in fundraising efforts:

The fall Brew Fest is a lot of fun. We always released the calendars at that event and hosted a signing party. All the beer distributers provide samples. Another great annual event started in recent years is the spring Wine Pairing. Everybody loves that. Gives restaurants a chance to show off their specialties, and it's wonderful—a real success.

The fire department is lucky. You have people with all kinds of skills totally dedicated to providing whatever the fire company needs. Those committees do such an amazing job. KIVFD administrator Sandra Early deserves a lot of credit for helping with fundraising over the last few years. She's done so much hard work. Like many people since the beginning of the fire company, she works completely and tirelessly for the organization. You're lucky when you have people like that involved.

Terry's proud to be associated with such a long line of energetic fundraisers and passionate volunteers.

Part VI
TIME IS THE FIRE

Time is the fire in which we burn.
—Delmore Schwartz, American writer

The Future

After years of working within the limited space of its original Stevensville firehouse, the Kent Island Volunteer Fire Department was at a crossroads. The *Bay Times* of March 27, 2002, reported:

> *In its 55-year history the men and women of the Kent Island Volunteer Fire Department have made a lot of crucial decisions to ensure the safety of the community. But the single biggest financial decision since the creation of the department in 1947 came on March 1 when the group settled on 9.6 acres of property for a new firehouse, said KIVFD Vice President Jody Schulz.*

Don Lewis, KIVFD president during the building of the new firehouse, points out the long-term thinking that went into the move. "When we sat down to build this station and everything since, we looked fifty years down the road. What are we doing now that will last us fifty years? That's how we make decisions."

Lou Johnson, former department board chairman and a driving force behind the county's retirement fund for volunteer emergency responders, considers the long process that went into the effort:

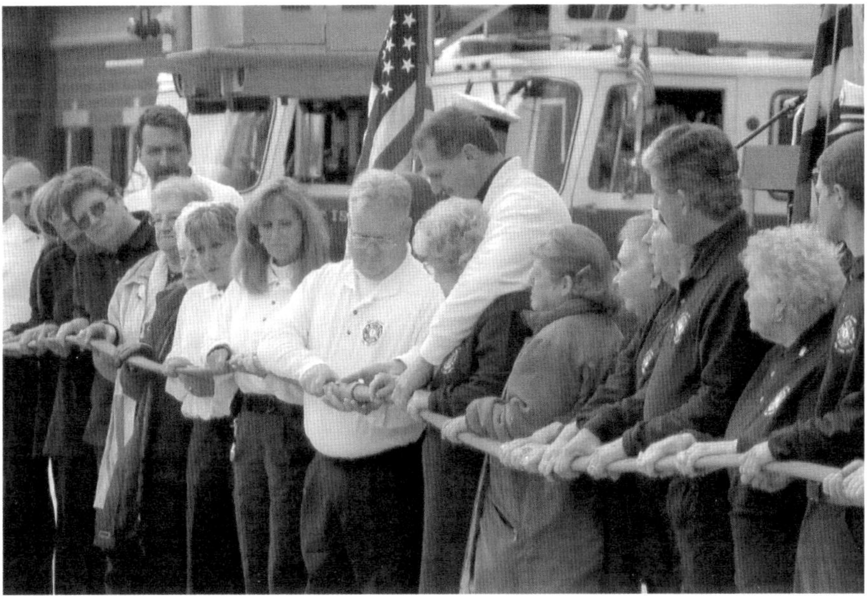

Members of the KIVFD and ladies auxiliary join together for the traditional uncoupling of the fire hose during the dedication ceremony of the new firehouse. *Courtesy Sandra Early.*

> *We spent ten years looking for property and thinking about what we were going to build. I went out as president in '99. We had a building committee working then. The next three presidents—Les Adams, Don Lewis and Francis Roudiez—were involved. Paul Schlotterbeck, Dave Reid, Woody Colburn and others toured the East Coast, from upper New York State to South Carolina, looking at new fire stations. Discussed what we liked about this one, what we didn't about another, then meshed it all together.*
>
> *We originally wanted to be behind the Acme at Tanner's Farm when this site became available. We knew we didn't need all this property; many options were considered as to what we could do with the rest of it. When Anne Arundel Medical Center decided to buy a piece from us, that was perfect. They brought in water and sewer lines, provided parking, even the road coming in. They paid for all that.*

The move was nevertheless expensive, as are the costs to operate such an organization. The numbers would stagger those founding volunteers. Chairman of finance and budget Doug Leatherman says, "This has become big business. You're sitting in a state-of-the-art $7 million firehouse that costs us $300,000 to 500,000 a year to run."

Todd Kerchner explains:

> *The fire company has always been treated as a business, but now it's a multimillion-dollar business. You have to deal with equipment issues, personnel issues, insurance and politics. The business is always changing.*

He adds a thought that members of the department strive to constantly remind their community: "The only thing that hasn't ever changed is the business is operated by volunteers."

Modern responsibilities include being more proactively involved with community issues that affect department operations. Attending county meetings and providing input on ordinances and site plan review for new development are now an important part of the job.

About 40 percent of costs are provided by the county; the remainder of the operating budget comes entirely from volunteer fundraising. Tom Balderson opines simply, "You think taxes are high now? They'd be astronomical without fundraising."

Doug Leatherman is unabashed about fundraising. "We get money everywhere we can. I need your money."

Tom cracks, "That's the finance and budget man speaking."

Doug continues, "But the public holds us accountable to spend their dollars wisely. This department is very conscientious of that. We can show every dollar raised, justify every dollar spent."

Lou Johnson says:

> *It'll be a long time coming, but eventually we'll have all paid firemen. In the meantime, it's a constant battle to convince residents who have moved here from more metropolitan areas that we're volunteers. Where they come from, the fire department is paid for with their taxes. Here that's not the case; here we're volunteers. Though it might feel like it sometimes, nobody lives at this firehouse. We respond from home, or work or wherever we are.*

And they're volunteers confronting new challenges in an ever-changing world. With a mission to evolve into a more proactive team, they have to keep up with advancements in technology and methodology. Retired firefighter Curtie Chance said, "You can't fight a 2010 fire with 1920 equipment."

In the company's early years, weeks might have gone by without an emergency call. In 2009, Station 1 responded to 1,166 calls—the most of any fire department in Queen Anne's County. Seventy-eight of those calls were fire or fire related, 714 were rescue or emergency service runs, 216 were considered "good intent" calls or unintentional false alarms, 21 were boat calls and 13 others were unspecified responses. Station 1 volunteers see themselves as leaders in a situation that will eventually affect their Eastern Shore colleagues. Former chairman of the Queen Anne's County Fire and Emergency Services Commission, Link George, believes that "because of all the growth here on the Shore, all local companies are going to face the changes we have."

The level of the volunteers' accountability has also changed. Doug Leatherman puts it this way:

> *You cover yourself and the department with every move. Used to be you'd go to a parade and have a beer or two, then go fight fire if you had to. You're not getting in that apparatus now if you've had any beer. If something happens while you're representing our department, you're going right to the doctor for urinalysis. Not just to find out if you're to blame, but also to clear you.*

"We're in a unique situation," summarizes Chief Paul Schlotterbeck.

> *Calling us a department or company might be a little misleading, but it's traditional. We're a quasi branch of county government, a private nonprofit corporation providing a public service. A volunteer organization that works with the county Department of Emergency Services and who ultimately answers to the county commissioners. It's not the good old boy days. Every call is scrutinized.*

EMS assistant chief Jeff Reiss oversees the ambulance service, which includes unpaid volunteers and, since 2008, two-per-day paid emergency medical technicians. "We're basically volunteers performing professional duties," Jeff says, "and performance review is done every time we leave our firehouse. Standards have to be met."

"It isn't as fun anymore, but it's still gratifying," observes twenty-one-year member Doug Leatherman, "when you see the look on a person's face when you've saved someone they love."

Tom Balderson knows what Doug means. "Sometimes you're the first person they see when their world is falling apart. You see how relieved they are someone is there. The first time you recognize that, it kind of throws you."

Doug acknowledges that the feeling's even stronger when thinking about the actions of others:

> *I never personally feel like a hero. I just do what somebody taught me to do, my job. But this past winter* [2009–10] *the department had an ice rescue with two little girls. I was so proud of my guys, I was at work and showed all my employees the newspaper and said, "Look at this." I was more proud for them than anything I've ever done myself. And I've been involved in some intense rescues. Saving people is dramatic. But it's important for everyone to know how much is involved with running a fire station. The members who do it are truly dedicated. They give countless hours in training, in working, in fundraising.*

"And we're your neighbors," adds Tom.

The Kent Island volunteers also note that they've become proactive with regards to educating the public. One of the most prominent components in this effort is a recently acquired Life Safety Trailer, a mobile unit where demonstrators can simulate various household dangers. Chief Paul Schlotterbeck supports increasing prevention awareness within the community because "the easiest fire to put out is the one that never starts."

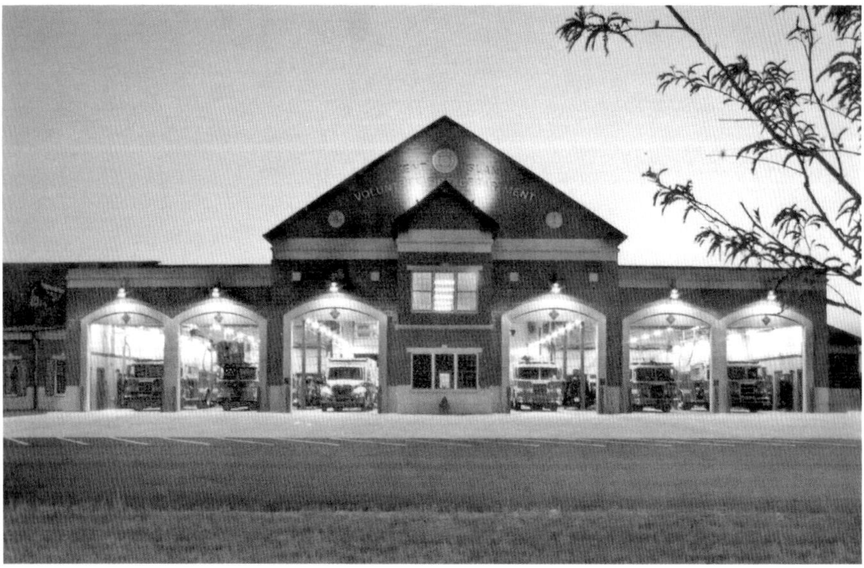

The Kent Island Volunteer Fire Department moved into its new Chester station on December 8, 2007. *Courtesy AsiPhotos.com/Stan Ruddie.*

Early member Bobby Norris recalls the first leaders of the KIVFD. "If the men I knew who started the fire company came here today to this new firehouse, they'd be busting buttons off their shirts they'd be so proud."

Former president Link George commemorates those groundbreaking volunteers: "The men who built the first firehouse did what they were supposed to do. They put up a quality building that lasted sixty years."

Calvin Duncan, life member and chief for over a decade, looks around in awe. "I don't know how you can get any more future than this new facility."

Ray Baker, life member and president in 1957 and 1958, concurs. "I think what's been done here is amazing. From the very first moment I saw it, I said, 'You done good, guys.'"

Korean War vet, department elder and past chief Paul "Friend" Fuld admitts, however, "The only problem I have with this new firehouse is too many damned doors. I never know where I'm going."

Everybody in the room laughs. They know Paul is not kidding.

Fill My Calling and Give the Best in Me

In her speech commemorating the Kent Island Volunteer Fire Department's fiftieth anniversary in 1997, Dolores Timms told her audience:

> *We have to applaud our fire department's ability to adjust to the highs and lows of their jobs: from the tragedy of the electrocution of a child to the delight of bringing a new child into the world. From the loss of a friend's home from fire to the triumph of a successful carnival, which provides the much-needed equipment to better perform their jobs. From the anxiety of performing duties on ice-slicked roads, in violent thunderstorms, scorching heat, bitter cold, night or day, tired or not, to the pride of accepting that trophy for "Best-Appearing Company" after polishing every piece of apparatus to perfection and every uniform to A-plus. We applaud your dedication to hours of training and hours of cleaning the firehouse and equipment to keep it in top working order to better serve the community. We can't forget all our sons and daughters who joined the fire department and were trained and guided into not only firemanship but also good citizenship by compassionate and understanding officers.*

Bobby Norris bears witness: "You were proud to be part of this organization because of the men and women who unselfishly gave of their time, talent and resources to build the company standing today."

Calvin Duncan agrees that the success of the fire department is due to the caliber of those involved. "You get in a building and it's on fire; you know who's behind you. My best memories are of the guys I worked for, and when I became chief, I couldn't ask for better than the people who served under me."

"The Kent Island Volunteer Fire Department *is* the community," asserts Link George.

Bobby Norris continues, "I learned a lot of lessons about life that I've carried with me."

"I'm just glad I got to be part of it," proclaims Marc Grollman. "I don't know what growing up would have been like without it. It was part of life, part of who you were. Knowing you had the responsibility and courage to do things others couldn't gives you confidence and pride."

"We were always more than a fire company," reminisces Bobby Timms. "We were a family, closer than blood. We worked together, played together, grew up together."

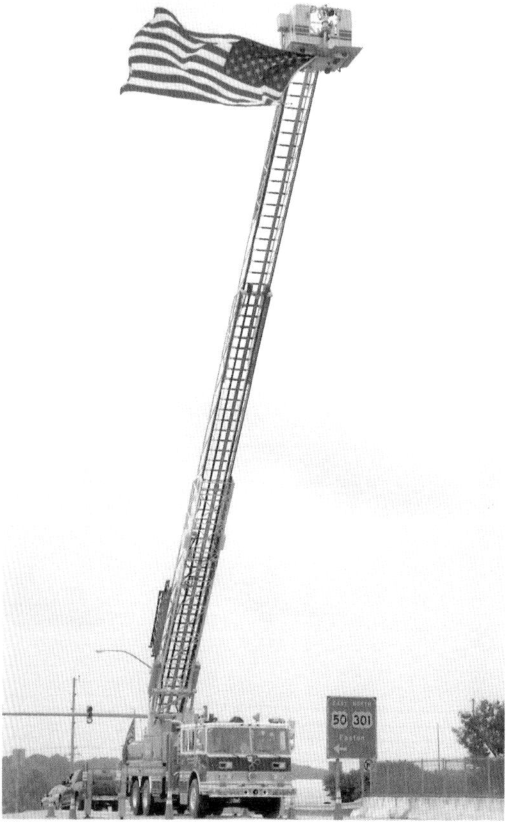

After the 9/11 terrorist attacks, the KIVFD let its patriotic spirit fly high on the Route 8 overpass. *Courtesy the* Capitol *newspaper/J. Henson.*

"No matter what happened—fires, accidents, storms, hurricanes, blizzards—we were and are there," Ray Baker affirms.

The last surviving founding member, Billy Lane, is by nature low-key. "I just felt it was something we needed, and we enjoyed helping out whenever we were needed."

Dave "Pops" Gellert, over seventy years old and still a top responder, appreciates what his service means to people. "It's nice when you're walking through a store and somebody says, 'Hey, there's the guy who saved Dad's life last week.'"

Lou Johnson can relate:

> *You get to know so many people. People will say, "Mr. Johnson, you saved my grandmother, and I really want to thank you guys." Makes you feel nine-feet tall. Everywhere we go people say, "Hi." Even though your role might be minute, it means a lot to others.*

"Just somebody saying thanks means a lot to you," says Calvin Duncan.

Ray Baker sums up his service as only a parent can:

> *When you're involved in an emergency, you're busy. You don't really think about things until it's over. I used to go home and look at my kids asleep in their bedrooms. Just stand there and look at them and be thankful.*

The conversation ends, and until the next alarm, the firehouse goes dark.

Appendix
RECIPES FROM STATION 1

Kent Island Crab Cakes

Love Point Hotel manager Hettie Fisher's early twentieth-century crab cake recipe provided by grandniece Nancy Sadler. Nancy was one of the first women to join KIVFD as an emergency responder.

1 pound crabmeat
1 egg
1 cup cornflakes
2 tablespoons mayonnaise
1 teaspoon mustard
¼ teaspoon salt
¼ teaspoon pepper
¼ teaspoon seafood seasoning

In a large bowl, fold together ingredients thoroughly. Pat into cakes. Brown in pan or on griddle until golden.

Firehouse Oyster Fritters

KIVFD Ladies Auxiliary icon Jackie Moore's original handwritten recipe provides ingredient measures for two different-sized batches. It does not, however, provide directions. Recipe provided by Jackie's daughter-in-law, Joyce Moore. Directions provided by author.

For 1 gallon of shucked oysters:
8 cups flour
12 teaspoons baking powder
8 eggs
5 cups milk

For 1 quart of shucked oysters:
2 cups flour
3 teaspoons baking powder
2 eggs
1¼ cups milk

In a large bowl, mix dry ingredients. Add wet ingredients and mix. Dip oysters in batter to coat. Fry in a large pan.

LADIES AUXILIARY CHICKEN SALAD DRESSING

This classic dressing was used in the chicken salad sandwiches the ladies auxiliary sold to motorists waiting to board the Matapeake Ferry in the late 1940s and early '50s. Great for large gatherings and meant to be served on homemade rolls, tripling the basic recipe will make 1 gallon of dressing or enough for 25 servings. Turkey can be used instead of chicken. Recipe provided by Dolores Timms.

2 eggs, well beaten
3 tablespoons vinegar
6 tablespoons sugar
2 cups mayonnaise (the ladies swear by Hellmann's brand)
Salt and pepper to taste

Combine eggs, vinegar and sugar in saucepan. Stir constantly over low heat until thickened. Let cool. Add mayonnaise, salt and pepper. Pour over 1 cooked, cooled and diced chicken and two stalks of celery, diced. Toss until evenly mixed.

For excellent potato salad, pour dressing over cooked and diced potatoes, diced celery, diced green pepper and diced onion. Toss with some celery seed and chill.

Mom Mom's Dinner Rolls

Madeline Lowery was one of KIVFD's first great cooks. This heirloom recipe was included in Seasoned with Love, *a family collection of Madeline's specialties dedicated to her memory. Note that scalding is a precise technique used to purify fresh milk and keep it from overthickening. Pasteurized milk does not require this step, though traditionalists might follow instructions as written, no questions asked. Recipe provided by Joyce Moore.*

1 yeast cake
1 cup warm water
4 tablespoons sugar
1 cup milk, scalded and cooled
1 egg, beaten
4–6 tablespoons shortening, softened
2 teaspoons salt
6 cups flour, sifted

Dissolve yeast and 1 tablespoon sugar in warm water, stir and let stand 10 minutes. Add milk, egg, shortening, remaining sugar, salt and 2½ cups flour. Beat smooth. Mix in remaining flour to form medium firm dough. Turn out on lightly floured board and knead until smooth. Place in greased bowl. Let rise until double size, about 1 hour.

Punch down dough, round up, cover and let rise until almost double, about 40 minutes. Fold down dough. Divide and shape into rolls. Heat oven to 400 degrees and bake 20 minutes or until golden brown.

Elaine Grollman's Applesauce Cake

A personal favorite of the author. Recipe provided by Elaine Grollman.

2 cups sugar
¾ cup margarine
2 cups applesauce
3 cups flour
2 teaspoons baking soda
1 teaspoon cinnamon
1 cup chopped walnuts
2 eggs

Cream together sugar and margarine. Add applesauce. Mix. Add flour. Beat. Add baking soda and cinnamon. Add nuts. Beat. Add eggs. Beat until mixed. Bake in Bundt pan or three loaf pans at 325 degrees for 40 to 55 minutes.

HOT MILK CAKE

Recipe from longtime ladies auxiliary member and KIVFD supporter Leona Clark Smith's kitchen provided by her daughter, Cindy Taylor.

4 eggs
2 teaspoons vanilla
2 cups sugar
2 cups flour
2 teaspoons baking powder
½ teaspoon salt
1 cup milk
1 stick butter

Beat eggs and vanilla until thick. Beat in sugar and flour. Mix together salt and baking powder. Heat milk and butter until butter melts. Add dry ingredients and milk to egg mixture. Beat 1 minute. Pour into greased and floured tube pan. Bake at 350 degrees for 45 to 50 minutes. Frost with chocolate icing.

COOKED PINEAPPLE CARNIVAL ICING

A cake wheel tradition provided by Joyce Moore.

1 small can crushed pineapple
1 stick butter
1 cup sugar with 3 heaping tablespoons flour added
Yellow food coloring

Bring pineapple and butter to a boil in a saucepan. Gradually add sugar/flour mixture until thickened. Add food coloring as desired. As mixture cools, it thickens. Use before completely set. Frosts one yellow cake.

BIBLIOGRAPHY

Emory, Frederic. *Queen Anne's County Maryland*. Baltimore: Maryland Historical Society, 1950.

Hale, Nathanial C. *Virginia Venturer*. Richmond, VA: Dietz Press, 1951.

Schoch, Mildred C. *Of History and Houses: A Kent Island Heritage*. Kent Island, MD: Kent Island Heritage Society, 1989.

Smith, Dennis. *History of Firefighting in America: 300 Years of Courage*. New York: Dial Press, 1978.

Tabler, Dan. *A Century of Service*. Centreville, MD: Goodwill Fire Company, Inc., 1989.

Truitt, Reginald V. *Kent Island, Maryland's Oldest Settlement.* Centreville, MD: Queen Anne's County Historical Society, 1964.

ABOUT THE AUTHOR

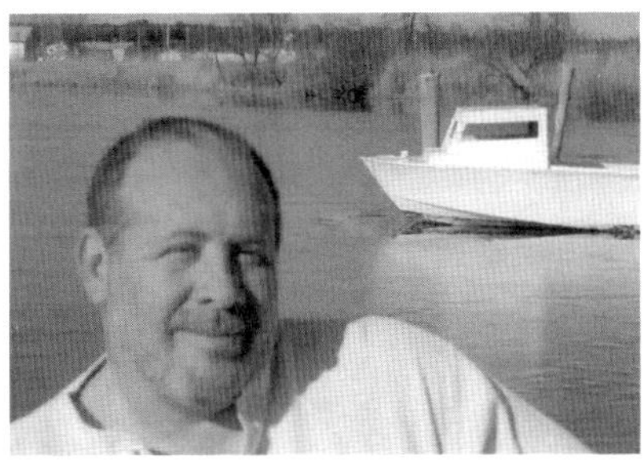

Brent Lewis is a native Eastern Shoreman. Brent's been appraising local residential real estate for over twenty years. He contributes regularly to The Kent Island Heritage Society newsletter and conducts interviews for the society's oral histories project. This is Brent's second book for The History Press. *Remembering Kent Island: Stories from the Chesapeake* was published in 2009.